Manufacturing, Heat Treatment and Forming of Advanced Metallic and Ceramic Materials

Editors

Marcin Wachowski
Henryk Paul
Sebastian Mróz

MDPI

Basel • Beijing • Wuhan • Barcelona • Belgrade • Novi Sad • Cluj • Manchester

Editors

Marcin Wachowski
Faculty of Mechanical Engineering
Military University of Technology
Warsaw
Poland

Henryk Paul
Department of Plastic Deformation of Metals
Polish Academy of Sciences
Cracov
Poland

Sebastian Mróz
Faculty of Production Engineering and Materials Technology
Czestochowa University of Technology
Czestochowa
Poland

Editorial Office

MDPI
St. Alban-Anlage 66
4052 Basel, Switzerland

This is a reprint of articles from the Special Issue published online in the open access journal *Materials* (ISSN 1996-1944) (available at: www.mdpi.com/journal/materials/special_issues/manufacturing_heat_treatment_forming_advanced_metallic_ceramic_materials).

For citation purposes, cite each article independently as indicated on the article page online and as indicated below:

Lastname, A.A.; Lastname, B.B. Article Title. *Journal Name* **Year**, *Volume Number*, Page Range.

ISBN 978-3-7258-0202-9 (Hbk)
ISBN 978-3-7258-0201-2 (PDF)
doi.org/10.3390/books978-3-7258-0201-2

Contents

Preface . . . vii

Sebastian Mróz, Karina Jagielska-Wiaderek, Piotr Szota, Andrzej Stefanik, Robert Kosturek and Marcin Wachowski
Effect of the Shape of Rolling Passes and the Temperature on the Corrosion Protection of the Mg/Al Bimetallic Bars
Reprinted from: *Materials* **2021**, *14*, 6926, doi:10.3390/ma14226926 . . . 1

Paweł Artur Król and Marcin Wachowski
Effect of Fire Temperature and Exposure Time on High-Strength Steel Bolts Microstructure and Residual Mechanical Properties
Reprinted from: *Materials* **2021**, *14*, 3116, doi:10.3390/ma14113116 . . . 18

Huu Chien Nguyen, Zdeněk Joska, Zdeněk Pokorný, Zbyněk Studený, Josef Sedlák and Josef Majerík et al.
Effect of Boron and Vanadium Addition on Friction-Wear Properties of the Coating AlCrN for Special Applications
Reprinted from: *Materials* **2021**, *14*, 4651, doi:10.3390/ma14164651 . . . 39

Janusz Kluczyński, Lucjan Śnieżek, Krzysztof Grzelak, Janusz Torzewski, Ireneusz Szachogłuchowicz and Artur Oziebło et al.
The Influence of Heat Treatment on Low Cycle Fatigue Properties of Selectively Laser Melted 316L Steel
Reprinted from: *Materials* **2020**, *13*, 5737, doi:10.3390/ma13245737 . . . 57

Jiri Prochazka, Zdenek Pokorny, Jozef Jasenak, Jozef Majerik and Vlastimil Neumann
Possibilities of the Utilization of Ferritic Nitrocarburizing on Case-Hardening Steels
Reprinted from: *Materials* **2021**, *14*, 3714, doi:10.3390/ma14133714 . . . 77

Michal Bumbalek, Zdenek Joska, Zdenek Pokorny, Josef Sedlak, Jozef Majerik and Vlastimil Neumann et al.
Cyclic Fatigue of Dental NiTi Instruments after Plasma Nitriding
Reprinted from: *Materials* **2021**, *14*, 2155, doi:10.3390/ma14092155 . . . 91

Marcin Małek, Waldemar Łasica, Marta Kadela, Janusz Kluczyński and Daniel Dudek
Physical and Mechanical Properties of Polypropylene Fibre-Reinforced Cement–Glass Composite
Reprinted from: *Materials* **2021**, *14*, 637, doi:10.3390/ma14030637 . . . 102

Rui Xu, Shijiao Zhao, Lei Nie, Changsheng Deng, Shaochang Hao and Xingyu Zhao et al.
Study on the Technology of Monodisperse Droplets by a High-Throughput and Instant-Mixing Droplet Microfluidic System
Reprinted from: *Materials* **2021**, *14*, 1263, doi:10.3390/ma14051263 . . . 121

Volodymyr Hutsaylyuk, Iaroslav Lytvynenko, Pavlo Maruschak, Volodymyr Dzyura, Georg Schnell and Hermann Seitz
A New Method for Modeling the Cyclic Structure of the Surface Microrelief of Titanium Alloy Ti6Al4V After Processing with Femtosecond Pulses
Reprinted from: *Materials* **2020**, *13*, 4983, doi:10.3390/ma13214983 . . . 136

Maciej Suliga, Radosław Wartacz and Marek Hawryluk
The Multi-Stage Drawing Process of Zinc-Coated Medium-Carbon Steel Wires in Conventional and Hydrodynamic Dies
Reprinted from: *Materials* **2020**, *13*, 4871, doi:10.3390/ma13214871 . . . . **144**

Preface

Metallic and ceramic materials play a significant role in the development of modern science and technology. They have a broad range of applications in various engineering fields. A proper manufacturing process, heat treatment, and forming are three main processes that can lead to obtaining metallic and ceramic materials which are created mainly to discover new or improved materials of construction, e.g., due to their strength properties. The main aim of this Special Issue was to publish original review and research articles from a wide range of research fields involving the manufacturing, heat treatment, and forming processes of metallic and ceramic materials.

Marcin Wachowski, Henryk Paul, and Sebastian Mróz
Editors

materials

MDPI

Article

Effect of the Shape of Rolling Passes and the Temperature on the Corrosion Protection of the Mg/Al Bimetallic Bars

Sebastian Mróz [1,*], Karina Jagielska-Wiaderek [1], Piotr Szota [1], Andrzej Stefanik [1], Robert Kosturek [2] and Marcin Wachowski [2]

1 Faculty of Production Engineering and Materials Technology, Czestochowa University of Technology, 42-201 Częstochowa, Poland; k.jagielska-wiaderek@pcz.pl (K.J.-W.); piotr.szota@pcz.pl (P.S.); andrzej.stefanik@pcz.pl (A.S.)

2 Faculty of Mechanical Engineering, Military University of Technology, 00-908 Warsaw, Poland; robert.kosturek@wat.edu.pl (R.K.); marcin.wachowski@wat.edu.pl (M.W.)

* Correspondence: sebastian.mroz@pcz.pl; Tel.: +48-692-401-124

Abstract: The paper presents the results of experimental tests of the rolling process of Mg/Al bimetallic bars in two systems of classic passes (horizontal oval-circle-horizontal oval-circle variant I) and modified (multi-radial horizontal oval-multi-radial vertical oval-multi-radial horizontal oval-circle-variant II). The feedstock in the form of round bimetallic bars with a diameter of 22 mm and 30% of the outer aluminum layer was made through explosive welding. The bimetallic bars consisted of an AZ31 magnesium core and a 1050A aluminum outer layer. Bars with a diameter of 17 mm were obtained as a result of rolling in four passes. The rolling process in the passes was conducted at two temperatures of 300 and 400 °C. Based on the analysis of the test results, it was found that the use of modified passes and a lower rolling temperature (300 °C) ensures a more homogenous distribution of the plating layer around the circumference of the core and results in an even grain decreasing, which improves the corrosion resistance of bimetallic bars compared to rolling bars in a classic system of passes and at a higher temperature (400 °C).

Keywords: Mg/Al bimetallic bars; explosive welding; groove rolling; microstructure; corrosion resistance

Citation: Mróz, S.; Jagielska-Wiaderek, K.; Szota, P.; Stefanik, A.; Kosturek, R.; Wachowski, M. Effect of the Shape of Rolling Passes and the Temperature on the Corrosion Protection of the Mg/Al Bimetallic Bars. *Materials* **2021**, *14*, 6926. https://doi.org/10.3390/ma14226926

Academic Editors: Frank Czerwinski and Maryam Tabrizian

Received: 27 September 2021
Accepted: 10 November 2021
Published: 16 November 2021

1. Introduction

Products made of magnesium alloys have been drawing more and more attention in many industries for over a dozen years. It relates to the low mass density and high strength of Mg alloys [1,2], while maintaining good plastic properties. As a result, such products are becoming more and more competitive to the widely used steel products. A significant impediment in the wider use of magnesium alloys in technology is their relatively poor corrosion resistance [3,4]. Thus, products made of magnesium alloys are, in many cases, covered with coatings, increasing their corrosion resistance [5,6]. One of many methods of securing products made of magnesium alloys is the application of aluminum as a protective outside layers [7,8]. Aluminum and its alloys exhibit considerably higher corrosion resistance than magnesium does. Thus, it can be expected that by producing a two-layered Mg/Al bar in the process of rolling in elongating has the advantages that both materials, magnesium and aluminum, could be combined. The tightness of the cladding layer (Al) and its appropriate uniform thickness on the perimeter and along the length seem to be the key to the effective inhibition of corrosion of the even more chemically active core material (an Mg alloy). The use of aluminum coatings or layers may be a prospective solution, ensuring an increase in the corrosion resistance of the magnesium core while not causing a significant increase in the weight density of the finished products. The advantage of using coatings of aluminum or its alloys is their excellent corrosion resistance in inert media as well as resistance to mechanical damage or abrasive wear. Due to the spontaneous passivation of aluminum and its alloys, unlike materials based on Fe or Mg, they are highly resistant to corrosion in a common, neutral air environment [9–11]. One of the methods ensuring the improvement of corrosion resistance of magnesium alloys is laser surface alloying [7,12] and thermo-chemical treatment [13], which are commonly used techniques for producing layers enriched in Al. The mentioned methods ensure the production of aluminum coatings with a maximum thickness of ~10 μm, which may result in a porous surface and deteriorate corrosion resistance [8]. Thus, it is advisable to search for methods of making aluminum layers of higher thickness, which will effectively protect the magnesium layer. Many works on the production of such layers on multi-layered plates, sheets, and round bimetallic Mg/Al bars can be found in the technical literature. The most commonly used methods include diffusion bonding [14,15], hot pressing [16,17], extrusion [18,19], rolling [20,21] and rolling in grooves [22,23], explosive welding [24,25], forging [26,27], and two-roll casting [28]. Additionally, in this case, some of the mentioned methods do not guarantee the right quality of the joint, the required thickness of the plating layer, and, in addition, in the case of round bimetallic bars, even distribution of the plating layer around the core perimeter [29,30]. The most frequently used methods to produce round bimetallic bars include extrusion [18,19,31] and rolling in grooves [23,30,32]. The feedstock for these processes is made directly in the processes themselves, as in the case of extrusion [18,19] or with the possibility of earlier use of the casting method [22] or explosive welding [30,32]. One of the methods ensuring the right thickness and even distribution on the perimeter of the round bar of the plating layer is the combination of the explosive welding method to produce a bimetallic feedstock, and then rolling in elongating grooves [32,33]. A combination of these methods was used to produce bimetallic bars of steel/Cu [29,34], Al/Cu [30], and, recently, also Mg/Al [22,23,32]. Although the method of explosion welding guarantees obtaining a bimetallic feedstock marked by a high quality of the joint [32], it does not always guarantee obtaining high-quality bimetallic bars while rolling in grooves. Due to the fact that the process of rolling bars in the grooves is marked by a spatial state of deformation and a large inhomogeneity of metal flow in the rolling gap, the distribution of deformations is inhomogeneous [29,34]. Additionally, while rolling bimetallic bars in the grooves, with an unfavorable ratio of the layer thickness to the core diameter, the difference in the plastic flow resistance of individual layers and poorly selected process parameters (deformation, temperature, shape of the grooves) may result in an uneven distribution of the thickness of the plating layer on the perimeter of the bar, and, in extreme cases, delamination of individual components [29,35].

In order to increase the homogeneity of deformation during the rolling of homogeneous bars [36–38], and recently also bimetallic bars [33], classic elongating grooves are subject to modification of their shape. In the earlier works of some authors [30,33] it was shown that using multi-radial modified elongating grooves in the circle-oval-circle system to obtain Al/Cu bimetallic bars, such a distribution of strains in individual components of rolled bimetallic bars can be obtained, which will ensure a significant increase in the evenness of the plating layer distribution around the perimeter of the core, with high quality of the joint at the same time. Such grooves were also used for rolling Mg/Al bimetallic bars. In one work [32] it was shown that the distribution of the aluminum layer thickness on the magnesium core was approx. 10% more even compared to Mg/Al bars rolled in the classic circle-oval-circle system. However, available literature data concerning the determination of the effect of process parameters on the pattern of Mg/Al bimetal flow and the possibility of its controlling during groove rolling are very scarce. Thus, the subject matter and scope of the paper will constitute a unique research output that will contribute to the development of a new group of bimetallic products of low specific gravity and enhanced corrosion resistance. The results obtained from the research will provide in the future a basis for carrying out studies within projects of an applied profile. The novelty of this word was to use the explosion welding method for the Mg/Al feedstock production and subsequent groove-rolling process using modified elongating grooves.

That is why the main purpose of using the modified grooves is to increase the uniform distribution of the outer Al layer to increase the corrosion resistance of Mg/Al bimetallic bars. In the existing work, there are no data on the impact of the shape of the grooves and process parameters on the corrosion resistance of Mg/Al bars. Thus, the novelty and the main aim of this work was to determine the impact of the shape of the grooves and the rolling temperature on the corrosion resistance of the bars by using an outer aluminum layer. As part of this research, the rolling process was conducted in two systems of elongating grooves: classic (horizontal oval-circle-horizontal oval-circle variant-I) and modified (multi-radial horizontal oval-multi-radial vertical oval-multi-radial horizontal oval-circle-variant II). The rolling process was conducted in four passes. The obtained Mg/Al bimetallic bars were subjected to complex tests of the layer thickness distribution around the perimeter, structural tests, microhardness, and corrosion resistance tests.

2. Materials and Methods

The bimetallic feedstock in the form of Mg/Al round bars used for rolling in grooves was obtained by the explosive welding method. The explosive welding process was conducted in cooperation with the Explomet company (Opole, Poland). The parameters of explosive welding have been described in detail in the previous works of some authors [24,32]. Eight sets of samples were prepared for testing, each consisting of aluminum tubes (grade 1050A) and magnesium bars (grade AZ31), respectively. Chemical composition of the materials used for the tests is given in Table 1. The diameter of the AZ31 bars was 19.2 mm. The outer diameter of the 1050A tubes was 24 mm and the wall thickness of the tube was 1.5 mm. The distance between the magnesium core and the inner diameter of the tube was 0.9 mm. A cylinder system was used for explosive welding [32].

Table 1. Chemical composition of the materials used for the tests [32] (reprinted with kind permission of Springer).

Material	Chemical Composition, % Mass.								
AZ31	Mg ball.	Mn 0.24	Cu –	Zn 0.72	Ca –	Al 2.8	Si 0.01	Fe 0.003	Ni 0.001
1050A	Al ball.	Fe 0.18	Cu 0.002	Mn 0.003	Mg 0.002	Zn 0.008	Ti 0.020	Si 0.06	Pb –

After explosive welding, the obtained Mg/Al bimetallic feedstocks were 400 mm long. Figure 1 shows the view of exemplary Mg/Al bimetallic feedstocks.

Figure 1. View of round Mg/Al bimetallic feedstocks after explosive welding.

The average diameter of the Mg/Al bimetallic feedstocks after explosion welding was 22.5 mm. The obtained bimetallic feedstocks were marked by a slight difference in the thickness of the outer aluminum layer around the perimeter of the magnesium core. The average thickness of the aluminum layer was 1.67 mm and its share in the cross-section of the bimetallic bar was 28% (Figure 2).

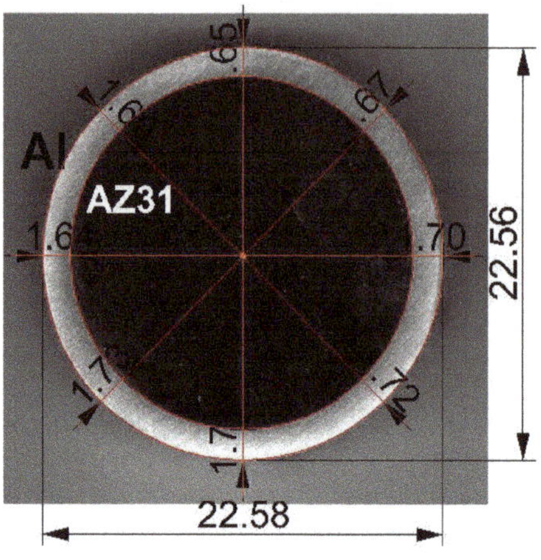

Figure 2. An exemplary shape of the Mg/Al bimetallic feedstock (cross-section) after explosive welding.

One of the deciding factors impacting the even distribution of the plating layer is the shape of the grooves and the distribution of deformations in individual passes. Two systems of elongating grooves were designed for the rolling process: the classic system of grooves, oval-circle-oval-circle (variant I), and its modification, multi-radial horizontal oval-multi-radial vertical oval-multi-radial horizontal oval-circle (variant II). In the case of both variants, the rolling process took place in four passes. The finished round pass was the same in both variants. The rolling feedstock was round bars with 22.5 mm diameter and 100 mm length. After rolling, the obtained round bars had a 17-mm diameter.

Computer simulations with the use of a computer program based on finite element method (FEM) were used to develop the design of the new passes. The assumption of using multiple radii in oval passes was to create the right conditions during deformation, limiting the controlled plastic flow of the magnesium core plating layer and, thus, obtaining a more even distribution of the plating layer around the perimeter of the magnesium core. The deformation pattern for individual variants is presented in Table 2. The shape of two developed groove systems is shown in Figure 3.

Table 2. Pattern of deformations used in the rolling process of Mg/Al bimetallic bars.

Rolling Variant	Coefficient of Elongation				Av. Coefficient of Elongation	Total Elongation
	Pass no. 1	Pass no. 2	Pass no. 3	Pass no. 4		
I	1.20	1.20	1.10	1.10	1.15	1.75
II	1.25	1.15	1.13	1.08	1.15	1.75

Temperature-deformation parameters were selected based on the results of physical and numerical modelings of compression tests of two-layer Mg/Al samples [39,40]. The rolling process was conducted for two temperatures: 300 and 400 °C. After each pass, the samples were reheated to the rolling temperature. The rolling speed was 0.2 m/s. A laboratory two-high rolling mill with a nominal diameter of working rolls of 150 mm was used for the experimental tests (Figure 4). After each pass, templets were collected to determine the thickness distribution of the plating layer around the perimeter of the magnesium core. The bimetallic feedstock was heated in the LAC KC 120/14 (LAC, Židlochovice, Czech Republic) resistance chamber furnace before rolling and before the individual passes.

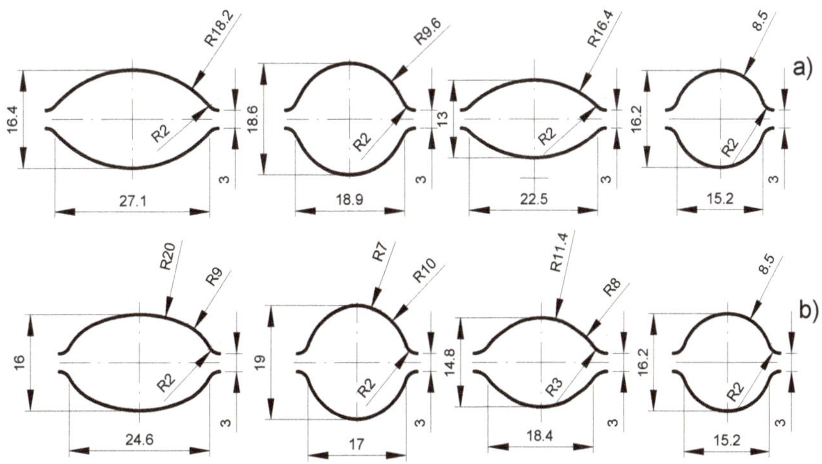

Figure 3. The shape and dimensions of the designed elongating passes: (**a**) classic system (variant I), (**b**) multi-radial modified system (variant II).

(**a**) (**b**)

Figure 4. Laboratory rolling mill D150 mm, general view (**a**); arrangement of the grooves along the width of the roll (**b**).

In order to determine the distribution of the plating layer for each sample collected after a particular pass, 32 measurements were conducted, which corresponded to the multiple of the angle of 11.25°, starting from the orientation (vertical symmetry axis of the groove) in the N direction. The evenness of the thickness distribution of the 1050A layer on the core of the AZ31 bar was determined using the coefficient of unevenness of the thickness distribution of the K_{plat} plating layer [40], defined as the ratio of the maximum thickness of the plating layer ($t_{\max}$) to the minimum thickness ($t_{\min}$), in the cross-section of the finished Mg/Al bimetallic bars, which can be determined using the following Equation (1):

$$K_{plat} = \frac{t_{\max}}{t_{\min}} \quad (1)$$

A detailed description of the methodology used to determine the analyzed coefficient K_{plat} is presented in [40,41]. In order to ensure the correctness of the tests, the aluminum layer thickness distribution was determined for three samples from batch bars and for three samples taken from corresponding rolled bars for the analyzed variants. The obtained difference in measurements for the charge after the explosion welding process did not exceed 5%, and for the rolled bars it was slightly higher, amounting to 8.6%. The increase in unevenness was closely related to the rolling operation and can be minimized by modifying the rolling equipment. This confirmed the correctness and repeatability of both the charge production processes and the finished bars used.

Microstructural tests of Mg/Al bimetallic bars included observations on an optical microscope Olympus LEXT OLS 4100 (Olympus, Tokyo, Japan) and a scanning microscope JEOL JSM-6610 (JEOL, Tokyo, Japan). The samples were included in the resin, sanded with 80, 320, 500, 800, 1200, 2400, and 4000 gradations, and then polished with a diamond paste of 3 mm and 1 mm. In order to reveal the microstructure of the AZ31 alloy, samples were etched with a reagent composed of 19 mL of ethanol, 2 mL of acetic acid, and 1 g of picric acid (digestion time of 30 s). Revealing the microstructure of 1050A was carried out using 1% HF with an etching time of 60 s.

To assess the impact of the groove shape and the rolling temperature on the corrosion resistance of Mg/Al bimetallic bars, potentiodynamic polarization curves were conducted in a 0.5M Na_2SO_4 solution acidified to pH = 4.0. The electrodes in the form of rotating disks were used for electrochemical tests, in which fragments of the side surfaces of the tested samples with an area of 0.2 cm^2 were used for the electrodes. All potentiodynamic tests were performed at the temperature of 25 ± 0.1 °C, with the rotational speed of the disc equal to 12 rpm^{-1} and the scanning speed of the potential of 0.005 V·s^{-1} using its shift from E_0 value of 0.3 V lower than E_{corr} to E_1 = +0.5 V (regarding AgCl/Ag). This type of methodology ensured the production of repeatable passive layers (in the cathode range, the natural, spontaneously formed oxide layers were reduced during surface preparation), and, on the other hand, it limited material consumption, thanks to fast scanning and turning off polarization at relatively low potential values. Each time before drawing the potentiodynamic curve, the test sample was held for 15 min in the corrosive solution, i.e., until the corrosion potential reached a stationary value.

3. Results and Discussion

The collected templets after individual passes were analyzed in terms of changes in shape and dimensions on the cross-section. Figure 5 shows that each pass, regardless of the used variant, was properly filled with a bimetallic flow. In any case, there was no overflow of the pass and no delamination was observed at the joint of the components. Rolling at temperature of 300 °C caused that the difference in the plastic flow resistance of the Al layer and the magnesium core decreased, which reduced the uncontrolled flow of the soft aluminum layer from the magnesium core [26]. As a result, the greater volume of the deformed flow moved more intensively in the longitudinal direction at the expense of reducing its width (widening). Thus, the width of the finished bars rolled at the temperature of 300 °C was smaller (Figure 5a,c) compared to the width of the bars rolled at the temperature of 400 °C (Figure 5b,d). The lower rolling temperature enabled us to obtain a greater coefficient in relation to the assumed rolling pattern (Table 2), which amounted to 1.85 elongation in variant II.

Figure 6 shows the average thickness of the plating layer for the feedstock after explosive welding and the finished Mg/Al bars after rolling with the use of two types of grooves.

The data presented in Figure 6 show that the average thickness of the plating layer varied depending on the rolling variant and the rolling temperature. According to the data in Figure 6, the lower rolling temperature resulted in a smaller spreading and greater elongation of the Mg/Al band. It influenced the limitation of the uncontrolled flow down of the Al layer in individual passes, thanks to which the average thickness of the plating layer was greater compared to bars rolled at the temperature of 400 °C, where we can observe higher flowing down of the cladding layer. The greater thickness of the plating layer may increase the corrosion resistance of the Mg/Al bars. The average thickness of the plating layer for the bars rolled at the temperature of 300 °C was higher by approx. 5% in relation to bars rolled at the temperature of 400 °C. The impact of the applied system of passes was much smaller in relation to the rolling temperature. For the system of modified grooves, regardless of the rolling temperature, the thickness of the plating layer was slightly lower than the thickness of the Al layer obtained for the system of classic passes. This difference was mainly due to the greater elongation of the bars rolled in the system of modified passes.

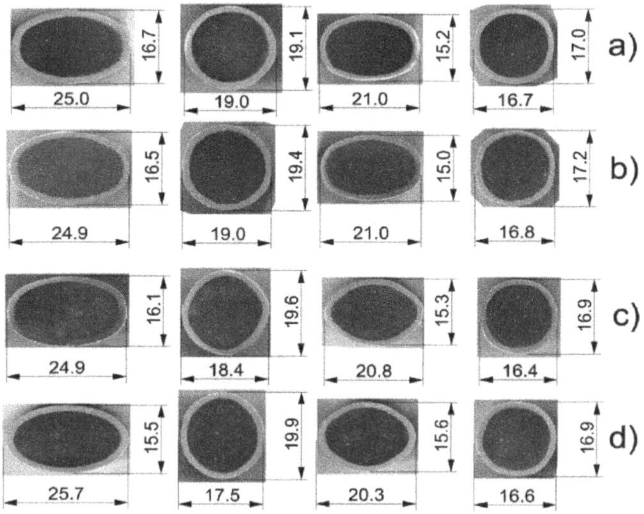

Figure 5. The shape and dimensions of the templets after rolling: (**a**) classic system of passes (variant I), rolling temperature of 300 °C; (**b**) classic system of passes (variant I), rolling temperature of 400 °C; (**c**) modified system of passes (variant II), temperature of 300 °C; (**d**) modified system of passes (variant II), temperature of 400 °C.

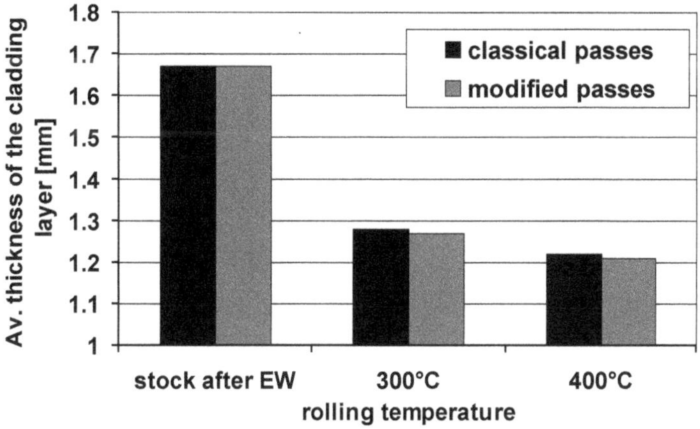

Figure 6. Average thickness of the plating layer around the perimeter of the magnesium core.

The main idea of using modified grooves (variant II) was to limit the uncontrolled plastic flow of the plating layer, thanks to which it will be possible to obtain Mg/Al bars with a more even distribution of the plating layer around the perimeter of the magnesium core. Limiting the local thinning of the plating layer on the perimeter and obtaining its even distribution is an important factor impacting the corrosion resistance of Mg/Al bars. Figure 7 presents the results of the calculations of K_{plat} coefficient calculated for finished bars rolled in two systems of passes.

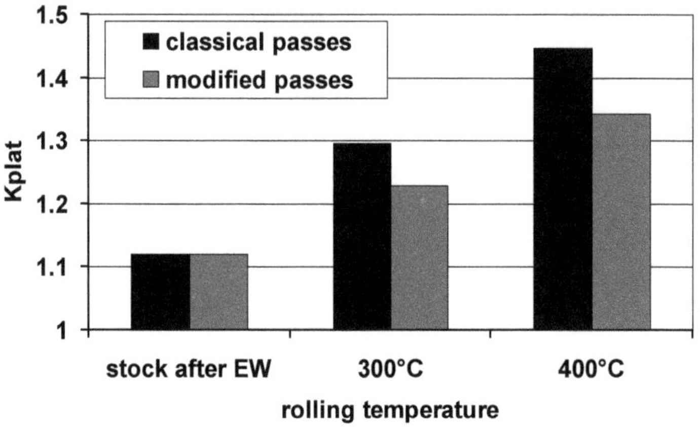

Figure 7. Values of K_{plat} coefficient depending on the rolling conditions.

The data presented in Figure 7 show that the developed new system of modified passes (variant II) had a positive effect on the thickness distribution of the Al layer. In each of the analyzed rolling variants, there was an increase in the unevenness of the plating layer distribution in relation to the feedstock after explosive welding; however, during rolling at the temperature of 300 °C with the use of modified grooves, the most even plating thickness distribution was obtained. For explosive welding, components in the form of AZ31 bars and Al tubes with perfect geometry were used. Thus, K_{plat} coefficient, marked by the uneven distribution of the plating layer, was close to 1 (K_{plat} = 1.12). Due to the fact that rolling in passes is marked by a spatial state of deformation and the lack of possibility to fully control the flow of the bimetallic band, it is natural that K_{plat} for bimetallic bars, after rolling in grooves, will increase in relation to the bimetallic feedstock. The modification of the shape of the passes, consisting of replacing the single-radius surface with multi-radial surfaces with straight sections, had a positive effect on the change in the deformation distribution and the evenness of the plastic flow of individual components in the Mg/Al bimK_{plat}etallic bar during rolling in the rolling gap (variant II). The new shape of the grooves enabled us to limit the accumulation of the plating layer material at the ends of the grooves towards the horizontal axis of the pattern symmetry and to reduce the thinning of the layer towards the bending at the same time. Thus, regardless of the rolling temperature, K_{plat} coefficient increased to a smaller extent in variant II in relation to Mg/Al bimetallic bars rolled in the classic system of passes (variant I). Rolling bars at the temperature of 300 °C, regardless of the applied rolling variant, had a positive effect on the evenness of deformation and plastic flow of the bimetal components. Thus, for Mg/Al bars rolled at 300 °C, K_{plat} coefficient was lower compared to the bars rolled at 400 °C. The most even distribution of the plating layer was obtained for Mg/Al bars rolled in passes modified for the rolling temperature of 300 °C. Detailed changes of K_{plat} coefficient for individual rolling variants are presented in Table 3.

Table 3. Change of K_{plat} coefficient in Mg/Al bars after rolling in grooves.

Temperature °C	Feedstock K_{plat}	Variant I K_{plat}	Change, % 2/3	Variant II, K_{plat}	Change, % 2/5	Change, % 3/5
1	2	3	4	5	6	7
300	1.12	1.34	19.6	1.23	9.8	−8.2
400	1.12	1.45	29.5	1.29	15.2	−11.0

The data analysis in Table 3 shows that the rolling of bars in the classic system of passes (variant I) increases the uneven distribution of the plating layer compared to the bimetallic feedstock by approx. 20% (temperature 300 °C) and by approx. 30% (temperature 400 °C).

The introduction of modified passes (variant II) significantly reduced the unevenness of the Al layer distribution on the perimeter of the Mg/Al bar. For the bars rolled at 300 °C, increasing K_{plat} coefficient in relation to the bimetallic feedstock totaled below 10%, and for the temperature of 400 °C, only 15%. Both values obtained for the rolled bars according to variant II, regardless of the rolling temperature used, were lower than the values obtained for bars rolled in the classic system of grooves (variant I). Comparing the obtained K_{plat} coefficients for individual variants, a greater evenness of the plating layer can be found for modified grooves, a reduction of K_{plat} by 8% for 300 °C and 11% for 400 °C, respectively.

Figures 8 and 9 show the results of microstructural tests for bimetallic bars rolled in classic and modified passes at the temperature of 300 °C and 400 °C, (cross-sections of samples).

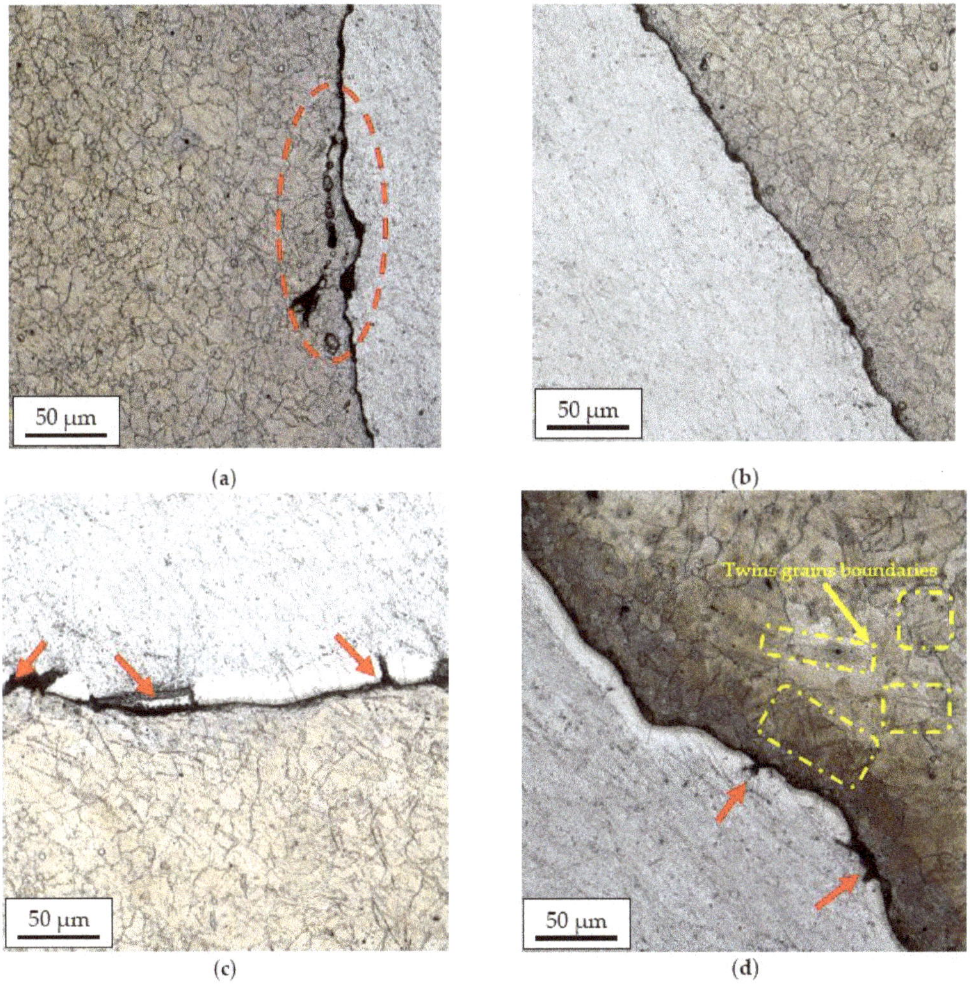

Figure 8. The AZ31 alloy microstructure after rolling in classic passes (variant I), temperature of 300 °C (**a**); modified passes (variant II), temperature of 300 °C (**b**); classic passes (variant I), temperature of 400 °C (**c**); modified passes (variant II), temperature of 400 °C (**d**).

Figure 9. The 1050A microstructure after rolling in classic passes (variant I), temperature of 300 °C (**a**); modified passes (variant II), temperature of 300 °C (**b**); classic passes (variant I), temperature of 400 °C (**c**); modified passes (variant II), temperature of 400 °C (**d**).

Based on the analysis of the microstructure, it can be concluded that bimetallic bars rolled at a lower temperature (Figure 8a,b) are marked by the presence of the finest grain of the AZ31 alloy in the joint area of approx. 10–15 μm. Despite the similar grain size, the joint in the sample taken from the bar rolled in the modified passes was free from imperfections, unlike the sample taken from the bar rolled in the classic passes, in which local delamination and cracks of the AZ31 alloy in the joint area were found (the area marked in red). Samples distorted at a higher temperature (Figure 8c,d) were characterized by both a larger grain size of the AZ31 alloy and the presence of a diffusion zone at the joint line with a visible presence of numerous imperfections in the form of cracks and delaminations (red arrows). In the case of a sample taken from the bar rolled in classic passes (Figure 8c), the differences were the least visible, and the microstructure itself was similar to the sample shown in

Figure 8a with the presence of a larger grain in the joint area (15–20 μm). In the case of a sample taken from the bar rolled in classic passes (Figure 8d), inhomogeneities in the granular structure were found, a fine-grained microstructure (approx. 10–15 μm) was noticed directly at the joint line, and there was an area with very large grains (40–50 μm) with a noticeable presence of twin boundaries (the area marked in red) behind this zone.

The microstructural analysis of the 1050A outer layers for most of the tested samples enabled the identification of large inhomogeneities in the grain sizes, regardless of the applied process parameters. The grain size ranged from 20 μm up to 200 μm. Most of the samples showed the dominance of the coarse-grained microstructure. The sample taken from the bar rolled in passes modified at the temperature of 300 °C (Figure 9b), in which the microstructure was much more distorted because of plastic working, clearly differed from this trend. The most damaged part of the material passes modified at the temperature of 300 °C (Figure 9b) was the central part of the 1050A alloy layer, where there were areas characterized by a relatively small grain size (5.0 ÷ 31.4 μm).

The implemented modification of the passes, consisting of substituting the single-radial pass surface with multi-radial surfaces with straight line segments, had the favorable effect of changing the strain, stress distribution, and the plastic flow of individual components in the Mg/Al bimetallic bar in the roll gap during rolling. It affected the higher decrease of the grain size. The modified roll pass design helped to limit the build-up of the cladding layer material and the ends of the passes in the direction of the axis of symmetry of the pass and, at the same time, to reduce the layer thinning in the rolling reduction direction. The new pass shape of passes reduced the non-uniform distribution of the deformation and stress value in the rolling direction. Additionally, lowering the rolling temperature to 300 °C had a favorable effect on the homogeneity of the deformation and plastic flow of the bimetal components and, consequently, the decreasing of the grain size. The most uniform distribution of the cladding layer and lower grain size was obtained for the modified passes for the rolling temperature of 300 °C.

More homogenous deformation in individual passes for the system of modified grooves and a lower rolling temperature resulted in a more homogenous grain size for the analyzed sample.

To determine the diffusion zone, the EDS analysis was performed for samples taken from rolled bars in two variants and at two temperatures, including a map of alloying elements on the AZ31–1050A joint surface (Figure 10). Microstructural observations carried out on a scanning electron microscope enabled us to identify the diffusion zone in individual samples. In the case of Al-Mg bars distorted at a temperature of 300 °C, a relatively small diffusion zone with a thickness of approx. 1–1.5 μm was observed, while the thickness of the layer in the sample obtained from the bar rolled in the modified passes (variant II) was slightly lower; however, due to the inaccuracy of the EDS method, it could not be confirmed at this stage of the research. As predicted, samples distorted at higher temperature (400 °C) were characterized by much thicker diffusion layers, which enabled their more detailed analysis. For each of the analyzed samples, the diffusion zone consisted of two layers with different shares of magnesium and aluminum. Earlier studies conducted by the authors on the deformation at high temperature of the Al-Mg joint indicated that the layers of the diffusion zone consisted of intermetallic phases, respectively: β (Mg_2Al_3) on the aluminum side and γ ($Mg_{17}Al_{12}$) on the magnesium side [42]. In the case of a sample taken from bars rolled in classic passes, the diffusion zone was approx. 13 μm thick with participation of individual phases of 10 μm Mg_2Al_3 and 3 μm $Mg_{17}Al_{12}$. In the sample taken from bars rolled in modified passes, the zone was slightly smaller and was characterized by a thickness of approx. 11 μm with the presence of 7 μm Mg_2Al_3 and 4 μm $Mg_{17}Al_{12}$.

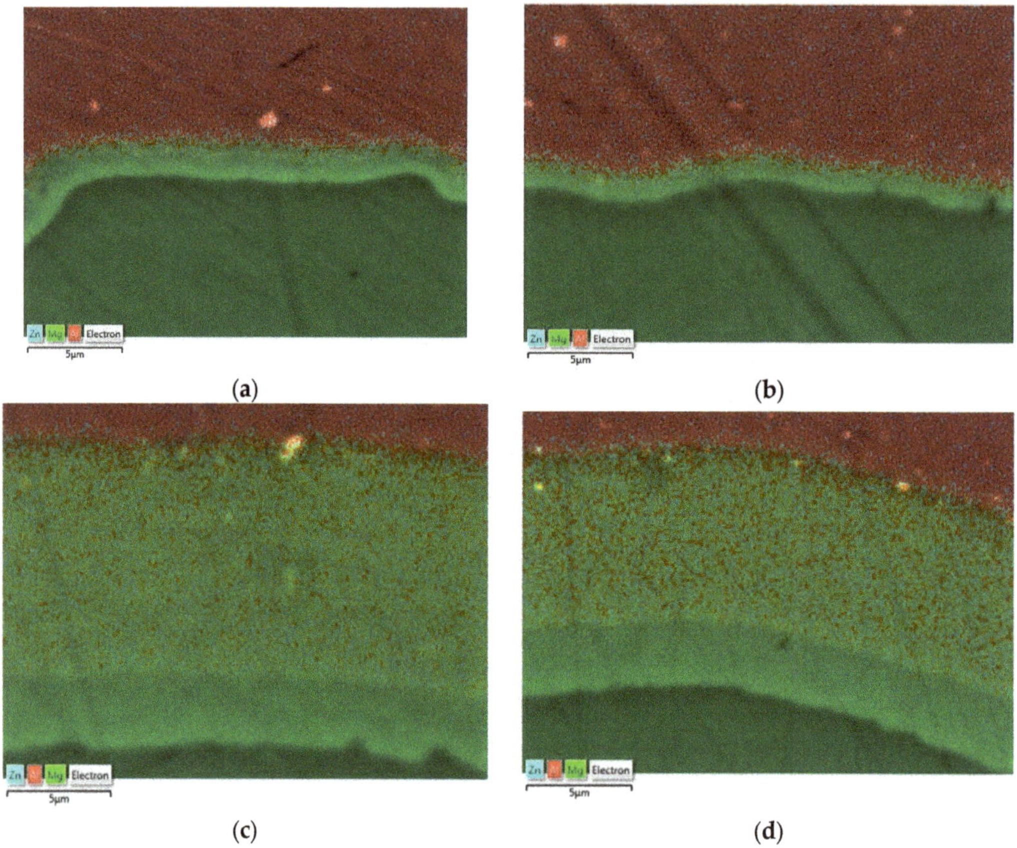

Figure 10. The EDS analysis of the joint zone: classic passes (variant I), temperature of 300 °C (**a**); modified passes (variant II), temperature of 300 °C (**b**); classic passes (variant I), temperature of 400 °C (**c**); modified passes (variant II), temperature of 400 °C (**d**).

The obtained polarization curves were drawn for material in their initial state (AZ31 alloy bar) and for Mg/Al bimetals rolled in two systems of elongating passes: classic (variant I) and modified (variant II) are shown in Figure 11. The feedstock materials in the form of AZ31 alloy bars and 1050A aluminum tubes used to produce bimetal, despite the small distance in the voltage series of metals (the normal potential for Mg is −2.37 V and for Al −1.66 V), showed definitely different corrosion parameters in an acidic environment [43,44]. As can be seen from the diagrams presented in Figure 11, in the applied corrosive environment, aluminum achieved anode currents more than three times the values lower than the magnesium alloy, which was the bimetal core. For the AZ31 alloy in corrosive solution with pH = 4.0, anode currents (i_a) reached values of approx. 270 mA/cm^2, while they did not exceed $1.4 \cdot 10^{-2}$ mA/cm^2 for aluminum. As it is known, in solutions with a pH of range from 4 to 8, in contrast to magnesium and its alloys, a permanent hydroxide was formed on the surface of aluminum, which was a passive layer, effectively protecting the surface of this material against corrosion [45].

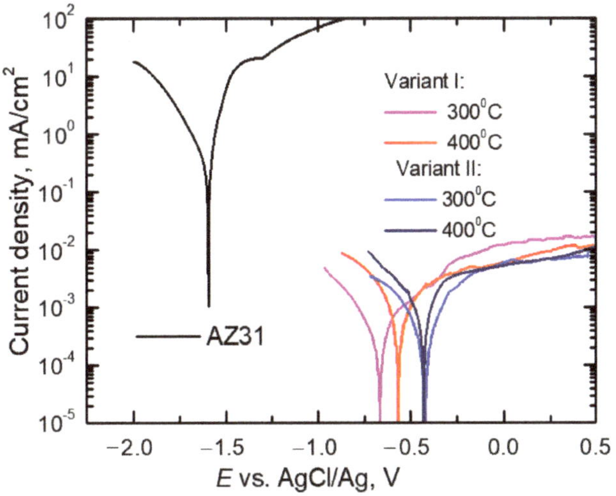

Figure 11. Potentiodynamic polarization curves for AZ31 and Mg/Al bimetals rolled at 300 °C or 400 °C for the classic (variant I) and modified system of grooves (variant II).

Comparing the schemes of the potentiodynamic curves of bimetals after rolling in the grooves, presented in Figure 11, it can be seen that their schemes were impacted by both the rolling method and the rolling temperature. As can be seen, the bimetals, after rolling in classic grooves (variant I), were marked by the lowest values of the corrosion potential (E_{corr}) and, thus, the earliest etching of the surface. In addition, the highest values of anode currents (i_a) were also found for this type of rolling. Modification of the rolling process (variant II) caused a significant shift of E_{corr} towards more positive values. Such an increase of E_{corr} was observed for samples after rolling according to variant II (system of modified passes), regardless of the temperature used in this process, and it proved that the processes of active etching of the surface were delayed. Precise values of E_{corr} and i_a read at E = 0.25 V are presented in Table 4.

Table 4. Parameters determining the corrosion resistance of the initial materials and the Mg/Al bimetal after various rolling variants.

Material/ Rolling Variant	Rolling Temperature [°C]	E_{corr}, [V]	i_a [mA/cm^2]	R_p [Ω·cm^2]	i_{corr} [mA/cm^2]
AZ31	-	−1.6	270	60	$860 \cdot 10^{-3}$
variant I	300	−0.66	$1.4 \cdot 10^{-2}$	$160 \cdot 10^{3}$	$0.3 \cdot 10^{-3}$
	400	−0.57	$1.0 \cdot 10^{-2}$	$50 \cdot 10^{3}$	$1.1 \cdot 10^{-3}$
variant II	300	−0.43	$0.7 \cdot 10^{-2}$	$160 \cdot 10^{3}$	$0.3 \cdot 10^{-3}$
	400	−0.43	$0.7 \cdot 10^{-2}$	$50 \cdot 10^{3}$	$1.1 \cdot 10^{-3}$

To determine the corrosion rate of Mg/Al bimetals, based on the drawn polarization curves, the polarization resistance (R_p), which is a parameter that determines the corrosion rate, was calculated. Figure 12 presents the relationships of linear polarization $\Delta E = E - E_{corr} = f(\Delta i)$ for equal potentials $E_{corr} \pm 20$ mV for bimetals after various rolling variants. For comparison, Figure 12 also present the same relationships for the starting materials from which the tested bimetals were made. As it is known, for potentials slightly different from the corrosion potential ($E_{corr} = \pm 20$ mV), i.e., in the range in which the Stern–Hoar equation [9] is fulfilled, the external current density is a linear function of the potential, and the slope of the corresponding lines is a measure of the polarization resistance (R_p)

inversely proportional to the corrosion rate (i_{corr}). Table 4 shows the characteristic values determined from the polarization curves that described the corrosion properties of the tested bimetal, i.e., the corrosion potential (E_{corr}), anode current (i_a), and corrosion current density (i_{corr}), determined based on the value of the polarization resistance (R_p) [46].

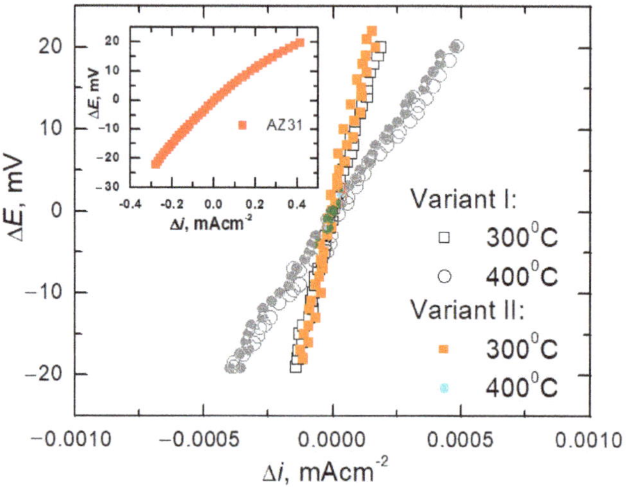

Figure 12. Measurements of linear polarization $\Delta E = E - E_{corr} = f(i_{outside})$ of input material (AZ31) and bimetals after classic (variant I) or modified (variant II) rolling at the temperature of 300 °C or 400 °C.

As can be seen from the data in Table 4, Mg/Al bimetals, after rolling processes, showed slightly more favorable corrosion parameters (lower values of i_{corr} and i_a) than Al, from which the coating was made. This slight improvement in the corrosion parameters of aluminum was probably due to the reduction of the grain size of the material due to the fragmentation of the primary grains during the deformation of the bars in individual passes. Similar results of an increase in the corrosion resistance of aluminum as a result of grain refinement were obtained after the forging (RS) [47] and angular pressing (ECAP) [48] processes. The authors attributed the increase in the corrosion resistance of the material to the formation of a compact and tight passive layer, the formation of which was fostered by both the high density of grain boundaries and dislocations. Detailed considerations on the influence of grain refinement on the corrosion resistance of aluminum are presented in [49]. While confirming the beneficial effect of grain grinding on the corrosion resistance of Al, the authors point out that the intensity of this influence depends on the corrosive environment. They also note that both the increase in the dislocation density and the associated stresses as well as the texture may affect the corrosion rate.

From a corrosive point of view, due to the slowest corrosion processes (low values of i_{corr}), the samples rolled at 300 °C are worthy of noticing. Regardless of the applied rolling variant, these bimetals corroded at the same rate, of $0.3 \cdot 10^{-3}$ mA/cm^2, while after rolling at 400 °C they corroded slightly faster ($i_{corr} = 1.1 \cdot 10^{-3}$ mA/cm^2). However, while the corrosion rates of the bimetals rolled at lower temperatures were the same, it should be noted that the more favorable values of both the corrosion potential and the anode currents were achieved by rolling with the modified method. The best corrosion characteristics obtained for Mg/Al bars rolled in modified passes for the rolling temperature of 300 °C probably reflect the fine-grained structure [49,50] and a more even distribution of the plating layer (Figure 7).

4. Conclusions

The use of modified passes in the circle-oval-circle system for rolling Mg/Al bimetallic bars results in the production of bars with a more even distribution of the plating layer around the core perimeter. A properly selected rolling temperature (300 °C) in combination with the applied deformation in the particular passes and the modification contributes to reaching bimetallic bars with high geometric accuracy of the cross-section. A more even deformation in the Al plating layer resulting from the deformation method and the reduced temperature results in an even grain fragmentation, which improves the corrosion resistance of bimetallic bars.

Rolling of Mg/Al bimetallic bars in modified passes at a temperature of 300 °C causes the smallest grain size of aluminum, which results in higher corrosion resistance of the final bars. In addition, it was found that bars rolled in modified passes are characterized by a slightly thinner diffusion zone and a lower presence of cracks and delamination in the joint area compared to bars rolled in a classic pass system, regardless of the rolling temperature.

The corrosion resistance of Mg/Al bimetallic bars is strictly dependent on both the temperature and the applied rolling method (the shape of the grooves). Conducting the rolling processes at the temperature of 300 °C slows down the etching processes of the bimetal surfaces, effectively reducing their corrosion rate. Moreover, the corrosion resistance of the Mg/Al bars is closely related to the thickness and uniform distribution of the plating layer, defined as the K_{plat} coefficient. The increase in the uniform distribution of the outer layer of Al as a result of the use of modified blanks amounted to approx. 10%, which had a direct impact on the general corrosion resistance of the entire Mg/Al system.

Among the applied rolling variants, bimetals rolled in modified passes are marked by better corrosion parameters, which is reflected in the shift of the corrosion potential towards the positive and lower anode currents. Higher E_{corr} values of the samples after modified rolling may prove that this treatment (compared to classic rolling) may additionally delay the start of etching of the bimetal surface in a corrosive environment.

The use of multi-radial modified elongating passes for rolling Mg/Al bimetallic bars, which were influenced by the plastic flow of the bimetal components and ultimately resulted in the finished bars with a uniform cladding layer distribution on the core, resulted in small dimensional deviations. Moreover, the determination of the effect of cladding layer thickness distribution and the cladding layer thickness on the magnesium core on the corrosion resistance of Mg/Al bimetallic bars has not been the subject of research so far.

Author Contributions: Conceptualization, S.M., K.J.-W., R.K. and M.W.; methodology, S.M., K.J.-W., P.S. and A.S.; formal analysis, S.M., K.J.-W. and R.K.; data curation, S.M., R.K., M.W., P.S. and A.S.; investigation, S.M., K.J.-W., R.K., M.W., A.S. and P.S.; writing—original draft preparation, S.M., K.J.-W. and R.K.; writing—review and editing, S.M. and A.S.; visualization, S.M., K.J.-W. and R.K. All authors have read and agreed to the published version of the manuscript.

Funding: Part of the research in the article was carried out with the use of funds by the National Scientific Centre (Poland) grant number DEC-2013/11/B/ST8/04352/1.

Institutional Review Board Statement: Not applicable.

Informed Consent Statement: Not applicable.

Data Availability Statement: The data presented in this study are available on request from the corresponding author.

Conflicts of Interest: The authors declare no conflict of interest.

References

1. Kaya, A.A. A Review on Developments in Magnesium Alloys. *Front. Mater.* **2020**, *7*. [CrossRef]
2. Rakshith, M.; Seenuvasaperumal, P. Review on the effect of different processing techniques on the microstructure and mechanical behaviour of AZ31 Magnesium alloy. *J. Magnes. Alloy.* **2021**, *15*, 1692–1714.
3. Zeng, R.; Zhang, J.; Huang, W.; Dietzel, W.; Kainer, K.U.; Blawert, C.; Wei, K.E. Review of studies on corrosion of magnesium alloys. *Trans. Nonferrous Met. Soc. China* **2006**, *2*, 763–771. [CrossRef]

4. Guo, K.W. A Review of Magnesium/Magnesium Alloys Corrosion. *Recent Pat. Corros. Sci.* **2011**, *1*, 72–90. [CrossRef]
5. Predko, P.; Rajnovic, D.; Grilli, M.; Postolnyi, B.; Zemcenkovs, V.; Rijkuris, G.; Pole, E.; Lisnanskis, M. Promising Methods for Corrosion Protection of Magnesium Alloys in the Case of Mg-Al, Mg-Mn-Ce and Mg-Zn-Zr: A Recent Progress Review. *Metals* **2021**, *11*, 1133. [CrossRef]
6. Gray, J.; Luan, B. Protective coatings on magnesium and its alloys—A critical review. *J. Alloys Compd.* **2002**, *336*, 88–113. [CrossRef]
7. Mola, R.; Dziadoń, A.; Jagielska-Wiaderek, K. Properties of Mg laser alloyed with Al or AlSi20. *Surf. Eng.* **2016**, *32*, 908–915. [CrossRef]
8. Mola, R. The properties of Mg protected by Al- and Al/Zn-enriched layers containing intermetallic phases. *J. Mater. Res.* **2015**, *30*, 3682–3691. [CrossRef]
9. Cottis, R.A.; Shreir, L.L. *Shreir's Corrosion*, 4th ed.; Elsevier: Amsterdam, The Netherlands, 2010.
10. Popov, B.N. *Corrosion Engineering—Principles and Solved Problems*; Elsevier: Amsterdam, The Netherlands, 2015.
11. Evans, U.R. *An Introduction to Metallic Corrosion*; Arnold, E., Ed.; Metals Park: London, UK; American Science Metals: Novelty OH, USA, 1982.
12. Mola, R.; Jagielska-Wiaderek, K. Formation of Al-enriched Surface Layers Through Reaction at the Mg-substrate/Al-powder Interface. *Surf. Interface Anal.* **2014**, *46*, 577–580. [CrossRef]
13. Ignat, S.; Sallamand, P.; Grevey, D.; Lambertin, M. Magnesium alloys laser (Nd:YAG) cladding and alloying with side injection of aluminium powder. *Appl. Surf. Sci.* **2004**, *225*, 124–134. [CrossRef]
14. Mola, R.; Cieślik, M.; Odo, K. Characteristics of AlSi11-AM60 bimetallic joints produced by diffusion bonding. *IOP Conf. Ser. Mater. Sci. Eng.* **2018**, *461*, 012058. [CrossRef]
15. Mola, R.; Mroz, S.; Szota, P. Effects of the process parameters on the formability of the intermetallic zone in two-layer Mg/Al materials. *Arch. Civ. Mech. Eng.* **2018**, *18*, 1401–1409. [CrossRef]
16. Li, X.; Liang, W.; Zhao, X.; Zhang, Y.; Fu, X.; Liu, F. Bonding of Mg and Al with Mg-Al eutectic alloy and its application in aluminum coating on magnesium. *J. Alloys Compd.* **2009**, *471*, 408–411. [CrossRef]
17. Zhu, B.; Liang, W.; Li, X. Interfacial microstructure, bonding strength and fracture of Magnesium-Aluminum laminated composite plates fabricated by direct hot pressing. *Mater. Sci. Eng. A* **2011**, *528*, 6584–6588. [CrossRef]
18. Tokunaga, T.; Szeliga, D.; Matsuura, K.; Ohno, M.; Pietrzyk, M. Sensitivity analysis for thickness uniformity of Al coating layer in extrusion of Mg/Al clad bar. *Int. J. Adv. Manuf. Technol.* **2015**, *80*, 507–513. [CrossRef]
19. Golovko, O.; Bieliaiev, S.M.; Nürnberger, F.; Danchenko, V.M. Extrusion of the bimetallic aluminium-magnesium rods and tubes. *Forsch. Im Ing.* **2015**, *79*, 17–27. [CrossRef]
20. Zhang, X.P.; Yang, T.H.; Castagne, S.; Wang, J.T. Microstructure; bonding strength and thickness ratio of Al/Mg/Al alloy laminated composites prepared by hot rolling. *Mater. Sci. Eng. A* **2011**, *528*, 1954–1960. [CrossRef]
21. Mroz, S.; Wierzba, A.; Stefanik, A.; Szota, P. Effect of Asymmetric Accumulative Roll-Bonding process on the Microstructure and Strength Evolution of the AA1050/AZ31/AA1050 Multilayered Composite Materials. *Materials* **2020**, *13*, 5401. [CrossRef] [PubMed]
22. Liu, N.; Chen, L.; Fu, Y.; Zhang, Y.; Tan, T.; Yin, F.; Liang, C. Interfacial characteristic of multi-pass caliber-rolled Mg/Al compound castings. *J. Mat. Proc. Technol.* **2019**, *267*, 196–204. [CrossRef]
23. Mróz, S.; Stefanik, A.; Szota, P. Groove rolling process of Mg/Al bimetallic bars. *Arch. Metall. Mater.* **2019**, *64*, 1067–1072.
24. Mróz, S.; Stradomski, G.; Dyja, H.; Galka, A. Using the explosive cladding method for production of Mg-Al bimetallic bars. *Arch. Civ. Mech. Eng.* **2015**, *15*, 317–323. [CrossRef]
25. Yan, Y.B.; Zhang, Z.W.; Shen, W.; Wang, J.H.; Zhang, L.K.; Chin, B.A. Microstructure and properties of magnesium AZ31B-aluminum 7075 explosively welded composite plate. *Mater. Sci. Eng. A* **2010**, *9*, 2241–2245. [CrossRef]
26. Mróz, S.; Gontarz, A.; Drozdowski, K.; Bala, H.; Szota, P. Forging of Mg/Al Bimetallic Handle Using Explosive Welded Feed-stock. *Arch. Civ. Mech. Eng.* **2018**, *18*, 401–412. [CrossRef]
27. Binotsch, C.; Nickel, D.; Feuerhack, A.; Awiszus, B. Forging of Al-Mg Compounds and Characterization of Interface. *Procedia Eng.* **2014**, *81*, 540–545. [CrossRef]
28. Bae, J.H.; Prasada Rao, A.K.; Kim, K.H.; Kim, N.J. Cladding of Mg alloy with Al by twin roll casting. *Scr. Mater.* **2011**, *64*, 836–839. [CrossRef]
29. Dyja, H.; Mróz, S.; Milenin, A. Theoretical and experimental analysis of the rolling process of bimetallic rods Cu-steel and Cu-Al. *J. Mater. Process. Technol.* **2004**, *153–154*, 100–107. [CrossRef]
30. Mróz, S.; Szota, P.; Stefanik, A.; Wąsek, S.; Stradomski, G. Analysis of Al-Cu bimetallic bars properties after explosive welding and rolling in modified passes. *Arch. Metall. Mater.* **2015**, *60*, 427–432. [CrossRef]
31. Priel, E.; Ungarish, Z.; Navi, N. Co-extrusion of a Mg/Al composite billet: A computational study validated by experiments. *J. Mater. Process. Technol.* **2016**, *236*, 103–113. [CrossRef]
32. Mróz, S.; Mola, R.; Szota, P.; Stefanik, A. Microstructure and properties of 1050A/AZ31 bimetallic bars produced by explosive cladding and subsequent groove rolling process. *Arch. Civ. Mech. Eng.* **2020**, *20*, 1–15. [CrossRef]
33. Mróz, S.; Szota, P.; Stefanik, A.; Wasek, S. 3D FEM modelling and experimental investigations of Al-Cu bimetallic bars rolling process in modified grooves. *Mater. Werkst.* **2015**, *46*, 300–310. [CrossRef]
34. Dyja, H.; Lesik, L.; Milenin, A.; Mróz, S. Theoretical and experimental analysis of stress and temperature distributions during the process of rolling bimetallic rods. *J. Mater. Process. Technol.* **2002**, *125–126*, 731–735. [CrossRef]

35. Golovanenko, S.A.; Meandrov, L.V. *Proizvodstvo Bimetallov*; Metalurgija: Moskva, Russia, 1966.
36. Abuda, E. Deformation during rolling in modified elongation groves. *J. Mat. Proc. Technol.* **1996**, *60*, 61–65.
37. Łabuda, E.; Dyja, H.; Korczak, P. Changes of pass geometry of the system of oval and vertical oval stretching passes in the rolling process. *J. Mater. Process. Technol.* **1998**, *80–81*, 361–364. [CrossRef]
38. Gronostajski, Z.; Pater, Z.; Madej, L.; Gontarz, A.; Lisiecki, L.; Lukaszek-Solek, A.; Łuksza, J.; Mróz, S.; Muskalski, Z.; Muzykiewicz, W.; et al. Re-cent development trends in metal forming. *Arch. Civ. Mech. Eng.* **2019**, *19*, 898–941. [CrossRef]
39. Mróz, S.; Szota, P.; Bajor, T.; Stefanik, A. Formability of Explosive Welded Mg/Al Bi-metallic. *Bar Key Eng. Mater.* **2016**, *716*, 114–120. [CrossRef]
40. Mroz, S.; Szota, P.; Bajor, T.; Stefanik, A. Theoretical and experimental analysis of formability of explosive welded Mg/Al bimetallic bars. *Arch. Metall. Mater.* **2017**, *62*, 501–507. [CrossRef]
41. Kazana, W.; Ciura, L.; Szala, J.; Hadasik, E. Measurements of geometrical parameters of clad wires. *Arch. Civ. Mech. Eng.* **2007**, *7*, 47–52. [CrossRef]
42. Wachowski, M.; Kosturek, R.; Śnieżek, L.; Mróz, S.; Stefanik, A.; Szota, P. The Effect of Post-Weld Hot-Rolling on the Properties of Explosively Welded Mg/Al/Ti Multilayer Composite. *Materials* **2020**, *13*, 1930. [CrossRef] [PubMed]
43. Surowska, B. *Wybrane Zagadnienia z Korozji i Ochrony Przed Korozją*; Wydawnictwo Politechniki Lubelskiej: Lublin, Poland, 2002.
44. Bialobrzeski, A.; Czekaj, E.; Heller, M. Corrosive properties of aluminum and magnesium alloys processed by pressure casting technology. *Arch. Foundry* **2002**, *2*, 294–313.
45. Pourbaix, M.; Staehle, R.W. *Lectures on Electrochemical Corrosion*; PWN: Warszawa, Poland, 1978.
46. Stern, M.; Geary, A.L. Electrochemical Polarization I. A Theoretical Analysis of the Shape of Polarization. *Curves J. Electro-Chem. Soc.* **1957**, *104*, 56–63. [CrossRef]
47. Abdulstaar, M.; Mhaede, M.; Wagner, L.; Wollmann, M. Corrosion behaviour of Al 1050 severely deformed by rotary swaging. *Mater. Des.* **2014**, *57*, 325–329. [CrossRef]
48. Song, D.; Ma, A.-B.; Jiang, J.-H.; Lin, P.-H.; Yang, D.-H. Corrosion behavior of ultra-fine grained industrial pure Al fabricated by ECAP. *Trans. Nonferrous Met. Soc. China* **2009**, *19*, 1065–1070. [CrossRef]
49. Ralston, K.; Fabijanic, D.; Birbilis, N. Effect of grain size on corrosion of high purity aluminium. *Electrochim. Acta* **2011**, *56*, 1729–1736. [CrossRef]
50. Ura-Bińczyk, E.; Bałkowiec, A.Z.; Mikołajczyk, Ł.; Lewandowska, M.; Kurzydłowski, K.J. The influence of grain size on the corrosion resistance of the 7475 aluminum alloy. *Ochr. Koroz.* **2011**, *2*, 44–47.

materials MDPI

Article

Effect of Fire Temperature and Exposure Time on High-Strength Steel Bolts Microstructure and Residual Mechanical Properties

Paweł Artur Król [1,*] and Marcin Wachowski [2]

1 Division of Concrete and Metal Structures, Faculty of Civil Engineering, Institute of Building Engineering, Warsaw University of Technology, 16 Armii Ludowej Ave., 00-637 Warsaw, Poland
2 Faculty of Mechanical Engineering, Military University of Technology, 2 Gen. S. Kaliskiego St., 00-908 Warsaw, Poland; marcin.wachowski@wat.edu.pl
* Correspondence: pawel.krol@pw.edu.pl; Tel.: +48-222-346-648

Abstract: In this study, the influence of different fire conditions on tempered 32CrB3 steel bolts of Grade 8.8 was investigated. In this research different temperatures, heating time, and cooling methods were correlated with the microstructure, hardness, and residual strength of the bolts. Chosen parameters of heat treatments correspond to simulated natural fire conditions that may occur in public facilities. Heat treated and unheated samples cut out from a series of tested bolts were subjected to microstructural tests using light microscopy (LM), scanning electron microscopy (SEM), energy dispersive spectroscopy (EDS), XRD phase analysis, and the quantitative analysis of the microstructure. The results of the microstructure tests were compared with the results of strength tests, including hardness and the ultimate residual tensile strength of the material (UTS) in the initial state and after the heat treatments. Results of the investigations revealed considerable microstructural changes in the bolt material as a result of exposing it to different fire conditions and cooling methods. A conducted comparative analysis also showed a significant effect of all such factors as the temperature level of the simulated fire, its duration, and the fire-fighting method on the mechanical properties of the bolts.

Keywords: effect of temperature; exposure time; steel microstructure; residual mechanical properties; high-strength steel bolts; heat treatment of steel; phase transformation; fire; cooling method

Citation: Król, P.A.; Wachowski, M. Effect of Fire Temperature and Exposure Time on High-Strength Steel Bolts Microstructure and Residual Mechanical Properties. *Materials* **2021**, *14*, 3116. https://doi.org/10.3390/ma14113116

Academic Editor: Adam Grajcar

Received: 7 May 2021
Accepted: 3 June 2021
Published: 6 June 2021

1. Introduction

The behaviour of steel structures during and after fire has stimulated the imagination of researchers and engineers for decades. Reflections derived from the observation of real fires in steel-structure buildings lead to the conclusion that disasters and collapses of these buildings most often happen not during the fire flashover, but during the fire extinction phase. When structural elements previously subjected to sufficiently long fire exposure lose or shorten their stability, this generates additional tensile forces that were not previously present. In such circumstances, internal forces are redistributed or connections between structural components are overloaded. At the same time, provided that the fire temperature has reached a sufficiently high level, microstructural changes can be seen in the structural materials. Observation of these structural destruction mechanisms permitted an assumption that high-strength bolts, which were subjected to heat treatment at the production process stage, lose their strength and load-bearing properties much faster than members made of conventional steel, thus determining the safety of steel constructions.

Brian Kirby is one of precursors of modern research on post-fire behaviour and residual mechanical properties of bolts. In [1,2], Kirby presented results of research carried out on M20-8.8 bolts made during various production processes—using hot and cold forging. He subjected the bolts to pre-heating in the temperature range of 20–800 °C, maintained in a given temperature level for 60 min, and then naturally cooled them in air. He observed that hot-forged bolts are more sensitive to temperature changes than cold-forged ones.

As a result of his research, he proposed using crystallographic methods to develop a method enabling identification of the maximum temperature that bolts could reach during a fire. He observed that post-fire residual strength of steel can be determined, inter alia, by specifying its hardness on a polished surface using the Vickers hardness tests with a load of 294.2 N (HV30) and comparing results obtained with those of a static tensile test, further bearing in mind that there is some orderly relationship between the hardness of steel and its tensile strength. He pointed out that if bolt production process parameters are known, this knowledge can be used to assess the temperature reached by a bolt during a fire by comparing its residual hardness with hardness of the material of the bolt in its initial state. At the same time, he demonstrated that if the fire temperature level exceeds the tempering temperature in the production process, the bolt material will soften. He showed that knowledge of metallurgical changes occurring in the bolt material can be helpful in diagnosing fire-damaged buildings.

The behaviour of bolts during a fire was also analysed, among others, by Gonzalez et al. [3] who focused their research on tensile tests of grade 10.9 bolts. Tests limited to destructive testing (static tensile test and static shear test) reflecting the behaviour of bolts under natural fire conditions (comprising the cooling phase) were also carried out by Hanus et al. [4]. They confirmed that if the structure is not destroyed during the temperature elevation phase, tensile forces generated in axially-restrained beams during the cooling phase may lead to the failure of bolted joints. Unfortunately, they did not conduct any research related to the analysis of material structure changes.

A number of works by Kodur et al. [5,6] are also worth paying greater attention to. In [5], the authors analyse the effect of temperature on variability of thermal and mechanical properties of high-strength steel bolts. They observed bolt failure mechanisms occurring during the tensile test through the prism of bolt microstructure that affected the material ductility and the failure model, as well as the shape of the fracture surface after failure. In [6], apart from routine destructive testing aimed at the assessment of residual mechanical properties of grade 8.8 bolts subjected to heating and controlled cooling cycles, they further devoted more attention to the issues of crack propagation and bolt failure models, conducting a broader analysis of fracture surface shapes obtained in the context of the target heating temperature level. They made an attempt to explain how cracks propagate depending on the microstructure of the bolt material. In [7], Yahyai et al. undertook similar research as in [6], focusing on bolts with a higher strength grade—10.9. The effects of a heating temperature, the chemical composition of the charge steel (raw material), and production process parameters on durable mechanical properties were subject to an in-depth analysis. A considerable part of the work was devoted to a detailed analysis on the surface of bolt failure, explaining the shape and form of fractures, and changes occurring in the steel microstructure.

It is worth mentioning that the available sources of knowledge present some research results relating only to two strength classes of bolts—grade 8.8 [1,2,4] and grade 10.9 [3,7]. The results obtained for bolts of grade 10.9 are qualitatively and quantitatively comparable to those presented for bolts of grade 8.8. Due to the fact that the bolts of strength classes lower than 8.8 are currently used only for connecting secondary elements, which are usually permanently deformed during a fire, it seems unreasonable to conduct fire tests using these bolts. It is assumed that such bolts will be replaced after a fire, including any damaged part to which they have been attached. In addition, due to the relatively low risk of a fire in a building during its technical life, so far—mostly for economic reasons—the widespread use of stainless steel bolts or special alloy steel bolts, including fire-resistant and creep-resistant ones, has not been adopted. These materials have a slightly different electrostatic potential, which would also require additional measures to prepare the contact surfaces of these bolts with structural steel elements in order to prevent accelerated galvanic corrosion. In current literature, however, one can find research devoted to fire tests of this type of bolts [8].

Some more interesting works from the borderline of material engineering and fire safety engineering, which are related to the issues discussed in this article, include the paper by Chi and Peng [9]; the authors present the results of steel plate tests that were heated to a temperature of 800 °C and higher and then quickly cooled in water. The work was aimed at demonstrating the possibility of reconstructing a fire scenario in a post-fire investigation on the basis of changes in the material microstructure in a situation where fire development and the level of the temperature reached in a fire were not known. The plates underwent, inter alia, metallographic tests to analyse their composition and microstructure and the results obtained were compared to the mechanical parameters of tested members destroyed by fire. The research included a detailed quantitative analysis of individual phase components. As a result of the heat treatment, the pearlite phase disappeared completely, ferrite was reduced from 80% to 30%, bainite increased its share to 30%, and martensite increased to 40%. A significant increase observed in the martensite phase changed the structure of the steel plate damaged by fire, which was reflected in the change of its material properties. Although yield point and tensile strength showed a growing trend, ductility of the member dropped significantly from 32.5% to 15%, which may result in a greater likelihood of sudden steel failure in structural elements of a building and translate into a decreased safety of its use.

Analyses of the effect of microstructure on mechanical properties of post-fire structural steels were also undertaken by Sajid et al. [10]. As part of the research, they analysed samples made from three grades of structural steel, subjected to heating in the temperature range from 500 to 1000 °C for 60 min, and then naturally cooled in the air. They presented changes in the microstructure in relation to the heating temperature. The analyses they conducted led to a conclusion that an increase in the share of ferrite fraction and the ferrite grain size leads to a decrease in the residual post-fire yield point and tensile strength, as well as an increase in the ductility of the tested structural steel grades. Based on the obtained results, the authors proposed multivariate linear regression equations to estimate post-fire/residual yield point of steel as a function of ferrite grain size and pearlite colony size. In their opinion, the results of these tests may be useful for future engineers and can be used to assess the quality and strength parameters of post-fire steel based solely on the results of microstructural tests, especially in situations where information of the temperature level reached in a fire is not available.

Works by Haiko et al. [11] and Xie et al. [12] are also worth mentioning. However, since they do not relate directly to the issues addressed in this article, they will not be broadly discussed.

The literature review shows that there is still a gap in the field of post-fire tests of steel structures devoted, in particular, to the selected components of connections and joints. In current published research, no attempt was made to determine the effect of the time duration of a developed fire or the applied fire-fighting strategy on the mechanical properties of the connectors, especially in close correlation with their microstructural changes. The aim of this paper is to fill this gap at least partially.

This article presents the results of research conducted to investigate the effect of various thermal and environmental conditions, typical for a real fire situation, on changes in the mechanical properties of high-strength construction bolts that results from their microstructural transformation. M20-8.8 bolts were used for the research for they are commonly used in prestressed butt joints and friction lap joints and even more commonly—due to their universal properties—used in regular non-prestressed joints. During the preliminary phase of testing, the bolts were subjected to thermal effects corresponding to selected conditions of a simulated fire, by heating them in batches in an electric furnace for the time specified in [13]. Following the exposure to high temperature, some bolts were set aside to cool down naturally. The intention was to recreate conditions of a spontaneous and natural fire, resulting either from a shortage of combustible substances or an insufficient amount of oxygen. The other part of the bolts was shock-cooled by immersion in water, which corresponded to a simulated firefighting operation carried out by rescue and firefighting teams. The time of heating corresponded to fire safety requirements adopted within the EU and established by law in relation to structural elements of buildings and building structures. Wording of national legal acts [13,14] directly implements the provisions of Regulation (EU) of the European Parliament and of the Council in 9 March 2011 [15], laying down harmonised conditions for the marketing of construction products. Their purpose is to ensure that basic requirements, such as required load-bearing capacity and structural stability, fire safety, health, and safety of use, are met.

2. Materials and Methods

Samples for microstructural tests were cut from a series of bolts previously subjected to fire exposure in various thermal conditions and, alternatively, also a simulated firefighting operation. The bolts were subjected to heating/tempering processes at various temperatures (400, 600, 800, and 1000 °C). In the case of the first series of samples, cooling was carried out naturally in the air (air-cooling/sample symbol: AC) by allowing the samples to cool down slowly, while in the case of the second series cooling was performed with an accelerated method which is immersion in water (water-cooling/sample symbol: WC). In each of the cases, two different heating times of 60 min and 240 min, respectively, were applied corresponding to the selected requirements resulting from [13]. A list of samples is presented in Table 1. Apart from the reference sample in the initial state (IS), the samples were labelled according to the X/Y/Z principle, where X denotes the cooling method, Y denotes the heating temperature, and Z denotes the heating time at a given temperature. Microstructural tests and hardness measurements were carried out using samples cut perpendicularly to the bolt shank axis, the area of which corresponded to the shank cross-section in the threadless place.

Table 1. List of tested samples.

Labels of Samples Prepared for Testing	List of Samples Tested with the Use of Light Microscopy	List of Samples Tested for the Purpose of the Quantitative Analysis
IS—reference sample	Yes	No
AC/400/60	No	No
AC/400/240	No	No
AC/600/60	Yes	No
AC/600/240	Yes	No
AC/800/60	Yes	Yes
AC/800/240	Yes	Yes
AC/1000/60	Yes	No
AC/1000/240	Yes	Yes
WC/400/60	No	No
WC/400/240	No	No
WC/600/60	Yes	No
WC/600/240	Yes	No
WC/800/60	Yes	Yes
WC/800/240	Yes	No
WC/1000/60	Yes	No
WC/1000/240	Yes	No

The bolts were made of an alloyed steel with an addition of boron, designated as 32CrB3, for which the chemical composition is provided according to the manufacturer's quality certificate in Table 2.

Table 2. Chemical composition of the 32CrB3 bolt steel according to the manufacturer's quality certificate [16].

Steel Designation	Chemical Composition [%]										
32CrB3	C	Mn	Si	P	S	Cr	Ni	Cu	Al	Mo	Sn
	0.31	0.84	0.13	0.012	0.013	0.74	0.08	0.15	0.025	0.018	0.010

In the production process, the bolts were made of smooth wire rod in the hot-forging process and then subjected to thermal improvement by quenching at a temperature of approximately 850–860 °C and tempering at a temperature of approximately 550 °C, which resulted in the obtaining of expected mechanical properties that corresponded to grade 8.8. Cut out samples for microstructural testing were hot-mounted in phenolic resin with Struers Multifast graphite filler and then ground using grinding wheels of 320, 600, 800, and 1200 gradation. The samples prepared in this manner were polished using a diamond suspension with a grain diameter of 3 µm and 1 µm. The microstructure of the steel was revealed by etching with a 2% solution of nitric acid in ethanol (2% NITAL).

In order to illustrate the microstructural changes resulting, inter alia, from phase transformations, the research material was analysed using the OLYMPUS LEXT OLS4100 digital light microscope and the JOEL scanning electron microscope (SEM), model JSM-6610, equipped with a secondary electron detector (SE) and a backscattered electron detector. The light microscopy tests were performed on the sample series shown in Table 1.

At the stage of microstructural tests, samples heated at the temperature of 400 °C were eliminated since the results of the static tensile test and the hardness test did not reveal any noticeable changes in relation to the initial state IS. This can be explained by the fact that this temperature is significantly lower than the tempering temperature used during the production process and the phase transformation temperature of steel, A_1. In the case of each of the samples, photos were taken at several points along the shank width, corresponding to their distance from the cross-sectional edge representing 1, 3, 5, 7, and 9 mm, respectively. For each of the points, two photos were taken at ×50 and ×100 magnifications, respectively, to obtain a more complete picture of changes occurring in the material microstructure. A scanning electron microscope (SEM) was used to test the initial state sample, with an accelerating voltage of 20 kV. The microscopic observations were supplemented with surface microanalysis of the chemical composition by means of the Oxford X-Max energy dispersive X-ray spectrometer (EDS), the results of which were generated in the form of an X-ray spectrum. In order to demonstrate the presence of carbides in the material structure, phase analysis was performed with the use of X-ray diffraction (XRD).

Images of the bolt microstructure in the form of photos obtained with the use of a digital light microscope at the magnification of ×100 were used for a quantitative analysis. Photos taken at a distance of 3 mm and 7 mm from the sample edges were selected for this purpose in order to capture differences, if any, in the microstructure along the width of the bolt shank. These images were used to outline grains of two basic phases, namely ferrite and pearlite, observed in the material microstructure. The quantitative analysis was performed using the MountainMap program. For each of the phases parameters such as the number of grains visible in the photo, their density, grain surface, or the mean equivalent grain diameter were determined. The types of samples utilized in this type of analysis are presented in Table 1. Considering the size of this article, results of the quantitative analysis were not presented in detail and only some of them are shown in Table 10 and Figure 1, which are included further in the work.

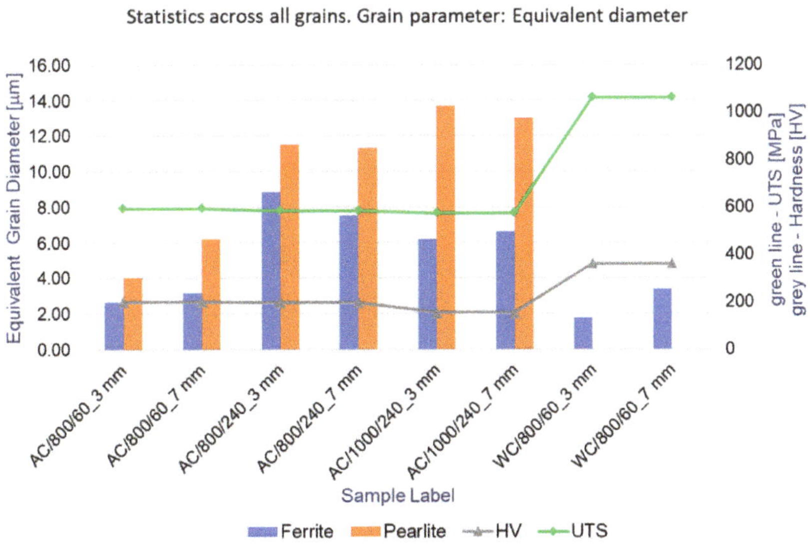

Figure 1. Diagram presenting dependence of microstructural indices, i.e., mean equivalent diameters of ferrite and pearlite grains, HV hardness, and ultimate tensile strength (UTS) on heat treatment parameters.

The Vickers HV hardness tests were carried out with a load of 294.2 N in the same places along the shank width and compared with the residual tensile strength values obtained from the previously performed static tensile test in order to confirm the correlation relationship between hardness HV and ultimate tensile strength UTS. For structural steels, it is presented in various literature sources in the form of the following relationship.

$$\text{UTS [MPa]} \approx (3.2 \div 3.5) \cdot \text{HV} \quad (1)$$

In order to eliminate human error, the hardness tests were carried out in an automated manner using the NEXUS 4300 stationary hardness tester in several places along the width of the bolt shank in order to capture any differences. Due to the lack of differentiation of the measurement results across the width of the sample, the mean value was taken as representative for further considerations.

3. Results

3.1. Microstructural Analysis

The analysis was aimed at investigating and assessing the effect of the secondary heat treatment resulting from exposure to various thermal conditions that may occur during a fire and the accompanying firefighting operation on the microstructure and mechanical properties of the bolt steel, which was previously quenched and tempered during production processes. The material in the initial state IS is characterised by a martensite structure (α phase—dark phase/areas in the photos of the IS sample) with a small quantity of residual austenite (γ phase—light areas in the photos of the IS sample) (Table 3).

Table 3. Microstructure of steel in its initial state.

Sample Label	Image at a Distance of 3 mm from the Sample Edge	Image at a Distance of 7 mm from the Sample Edge
IS—initial state (reference sample)	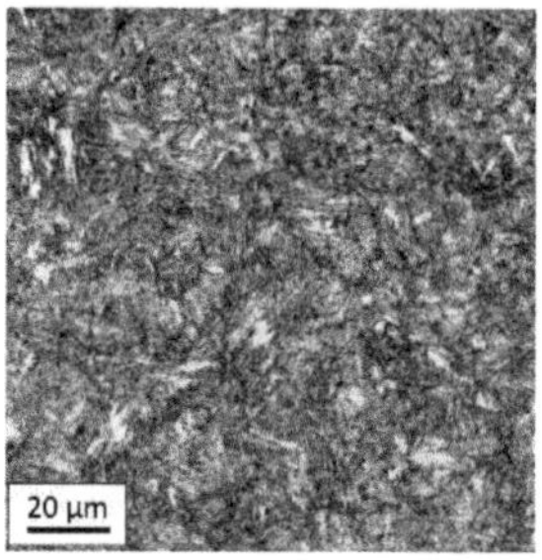	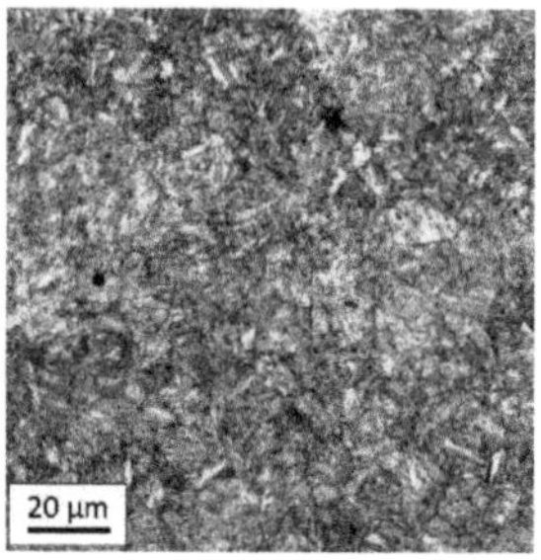

Analysis of the photos included in Table 3 does not show any significant differences between the microstructure image seen in the photo taken at a distance of 3 and 7 mm, respectively, from the shank edge. The presence of the martensite structure is related to the quenching process previously carried out, which is typical of steel bolt products and particularly those with increased strength. The martensite structure occurs throughout the cross-section of the tested element along the entire width of the bolt shank.

Due to the 0.3% carbon content, the steel covered by the research is referred to in the literature as hypoeutectoid steel and, in its unquenched state, is characterised by a ferrite and pearlite structure. Heat treatment of steel in the quenched state is called tempering and consists in heating up of steel to a temperature below the critical value A_1 = 727 °C (read out from the Fe-C phase diagram), leaving it at this temperature, and slowly cooling to ambient temperature. Generally, depending on the heating and cooling rate as well as the amount and type of alloying elements in low-alloy carbon steels, the critical temperature may somewhat vary [17]. Heat treatment below temperature A_1 does not lead to the formation of austenite (γ phase), while annealing above A_1 takes place when the ferrite (α phase) and austenite (γ phase) states coexist. Processes that occur during tempering are closely related to the phenomenon of diffusion of carbon and alloying elements, therefore they depend on both the temperature level and the heat treatment time. The purpose of the annealing process is to obtain a more fine-grained and more plastic structure, which is desirable in the context of more predictable behaviour and a non-brittle model of structural failure. Fine graining leads to an increase in the yield point since a denser mesh of grain boundaries is shaped in the structure of the material, which may slow down the formation of dislocations in the steel crystal lattice. According to the Hall–Petch formula (Equation (2)) [18], the yield point is inversely proportional to the square root of the mean grain size, which confirms the previously formulated thesis.

$$f_y \sim 1/(d)^{0.5} \tag{2}$$

As confirmed during our own lab-tests, the value of the yield point remains practically unchanged in the temperature range up to approximately 600 °C. In higher temperatures, grains start to grow, which reduces the density of mesh of grain boundaries, and leads to a reduced yield of steel. In order to confirm a chemical composition of steel (Table 2), an EDS analysis was performed using a scanning electron microscope. The analysis showed the presence of the following alloy elements: carbon, iron, manganese, and chromium. The content of manganese (1.0 wt.%) and chromium (0.9 wt.%), Table 4, turned out to be slightly higher than the content specified in the metallurgical certificate [16] (Table 2).

Table 4. Microstructure of steel in its initial state along with the results of the EDS analysis.

Microstructure of Steel in the Initial State	Results of the EDS Analysis
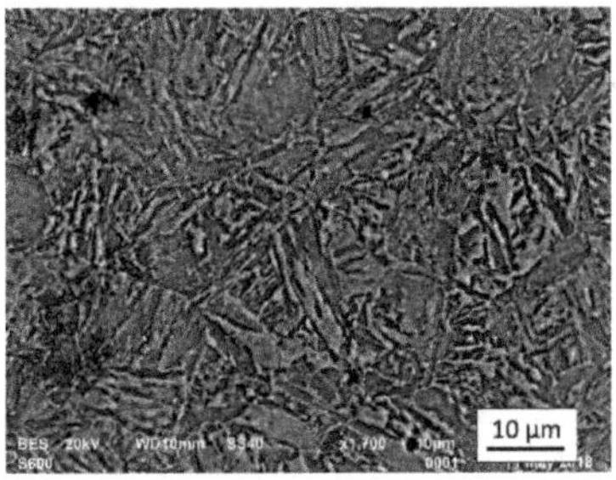	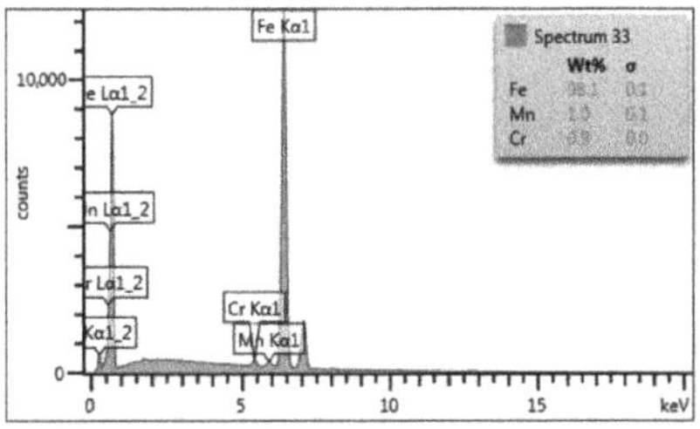

3.1.1. Microstructural Analysis of Samples Heated at 600 °C

After tempering at a temperature of 600 °C with both air-cooling and water-cooling methods, the martensitic microstructure of tempered steel is still preserved. Quantitative analysis of microstructure revealed no changes in size of martensite needles. This temperature, although it exceeds the nominal tempering temperature used during the production process of this grade of bolts, turned out to still be too low to initiate the phase transformation, consisting in the decomposition of martensite and the release of austenite, Table 5. Such an excess in comparison to the nominal tempering temperature is probably the result of the presence of alloying elements that may have caused certain disturbances in technological parameters in the case of repeated thermal treatment.

The analysis of the photos presented in Table 5 does not reveal any significant differences in the microstructure, neither along the width of the shank nor between the samples with different heating times or cooling methods.

Although there are no visual differences in the microstructure between the IS AC/600 and WC/600 samples, in the case of both the air-cooled and water-cooled bolts, the value of residual tensile strength after heating for 60 min turned out to be lower than the initial value of the IS bolts by approximately 12% and in the case of bolts heated for 240 min it was by as much as 22%. These tests showed the effect of the heating time on the value of residual strength properties of fire-exposed bolts.

3.1.2. Microstructural Analysis of Samples Heated at 800 °C

Tempering at the temperature of 800 °C with air cooling (AC/800/60 and AC/800/240 samples) leads to the martensite → austenite phase transformation. Then, during the cooling phase, ferrite is released (light areas) at the boundaries of austenite grains (Table 6). The temperature of 800 °C is above the A_1 line on the Fe-C phase diagram, therefore martensite austenitizes at this temperature.

Table 5. Microstructure of steel after heating at 600 °C for 60 min and 240 min with air-cooling and water-cooling.

Sample Label	Image at a Distance of 3 mm from the Sample Edge	Image at a Distance of 7 mm from the Sample Edge
AC/600/60	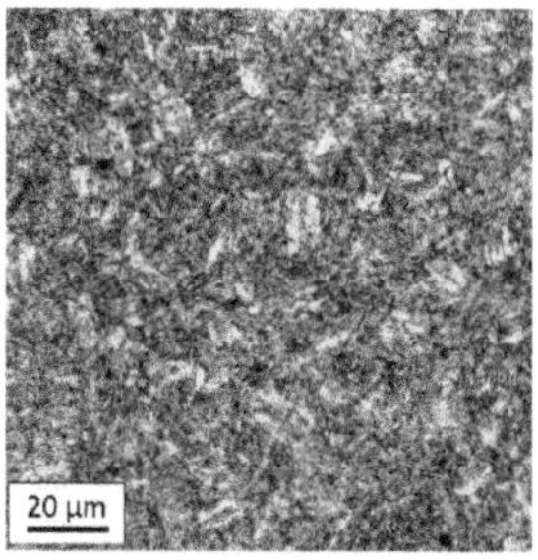	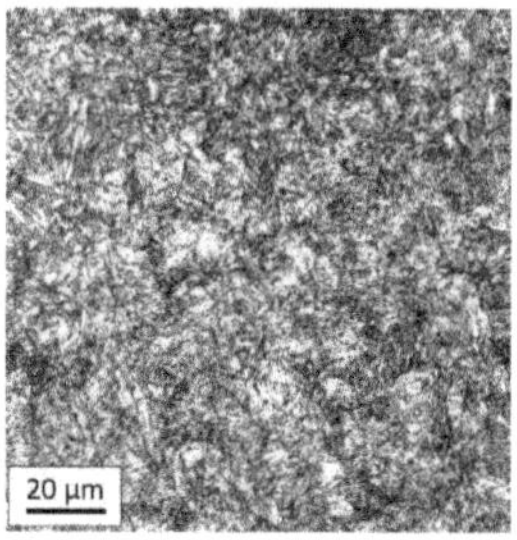
AC/600/240	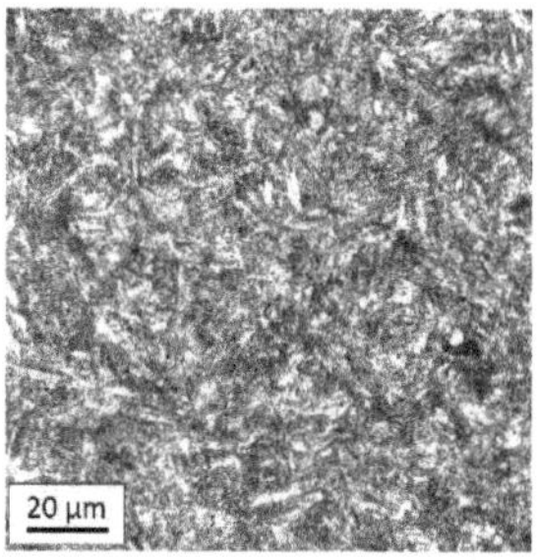	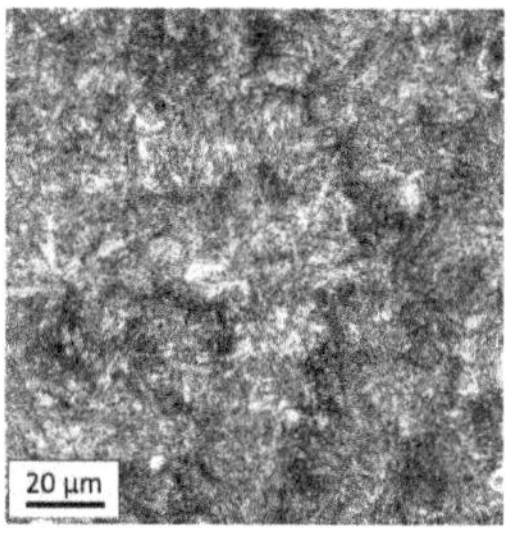
WC/600/60	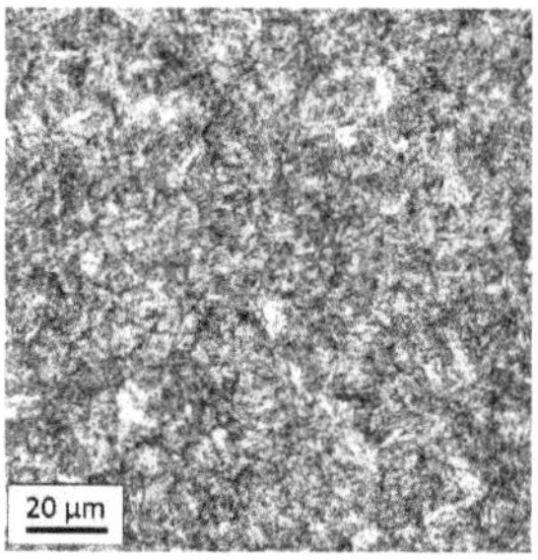	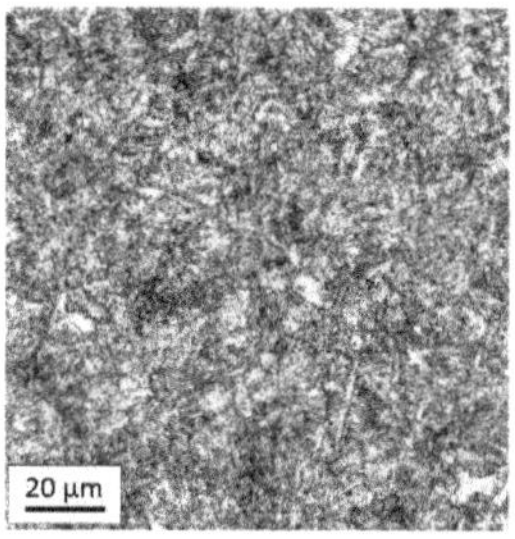
WC/600/240	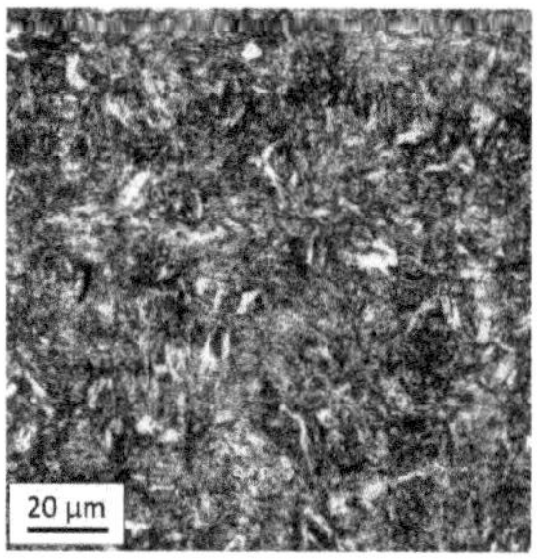	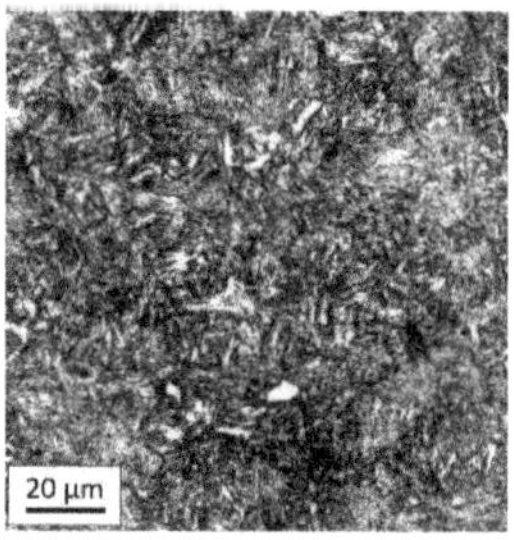

Table 6. Microstructure of steel after heating at 800 °C for 60 min and 240 min with air-cooling.

Sample Label	Image at a Distance of 3 mm from the Sample Edge	Image at a Distance of 7 mm from the Sample Edge
AC/800/60	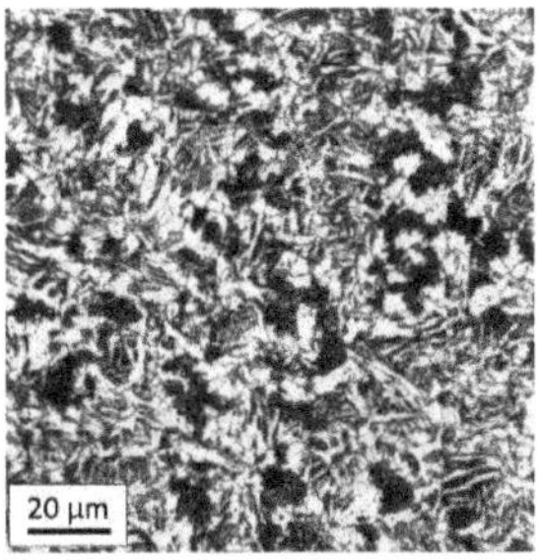	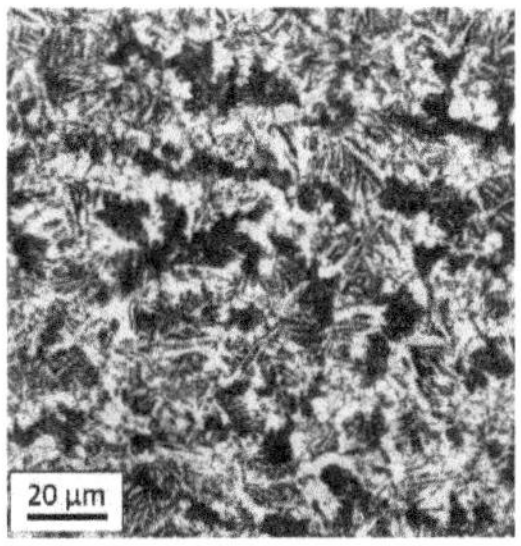
AC/800/240	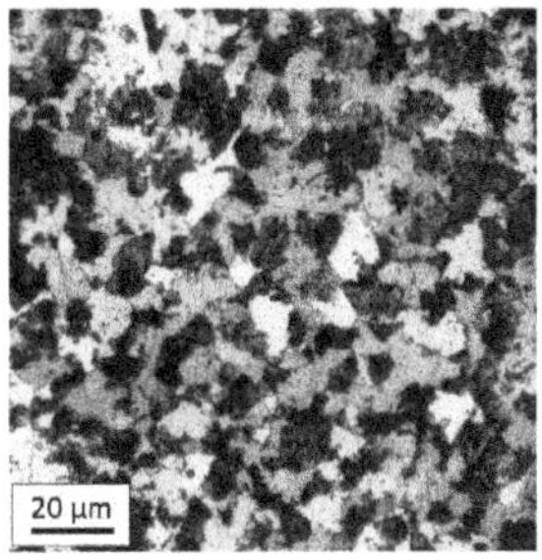	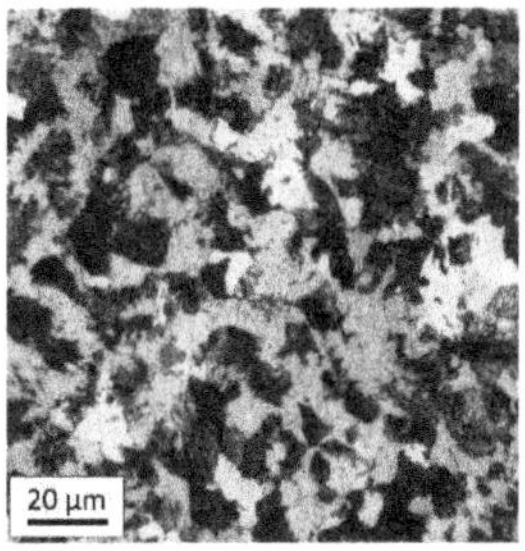

The austenite transformation during the tempering of alloy steels is influenced by the content of elements dissolved during austenitisation. Chromium, which is an alloying element of the tested steel, noticeably increases its transformation temperature and as a result the amount of ferrite released during the heating at 800 °C is small, but it increases along with growing temperature and heating time. When analysing the photos of the microstructure shown in Table 6, it can be easily observed that the microstructure of the steel heated at 800 °C for 240 min is quite different from that heated for 60 min only. The grain size increases, the mesh of boundaries between respective phases loosens, which results in a further reduction in the residual tensile strength and hardness of the material and thus becomes more plastic. Chromium, present in the tested steel as one of the alloying elements in the range up to 1%, slightly reduces the hardness of ferrite and considerably increases its impact strength along with a simultaneous decrease in hardness as compared to the reference value at room temperature (HV = 324). Moreover, it increases the amount of plastic residual austenite in the quenching process. Since the concentration of carbon in ferrite is lower than in the austenite from which is released, the carbon content in austenite increases along with heating. After reaching the critical value, i.e., after exceeding the limit level of the solubility of carbon in austenite, the pearlite transformation begins and leads to the transformation of the remaining austenite into pearlite (dark areas), which results in obtaining a plastic and soft ferrite–pearlite structure.

The transformation into ferrite–pearlite structure is demonstrated by a reduction in the residual ultimate tensile strength (UTS) from the initial state value equal to 1001 MPa (maximum tensile force F_m = 245 kN) to the level of UTS = 595 MPa (maximum tensile force F_m = 146 kN), with HV = 200 for the heating time of 60 min and UTS = 585 MPa (maximum tensile force F_m = 143 kN), and with HV = 196 for the heating time of 240 min. It is worth noting here that in the case of samples heated at the temperature of 800 °C and air-cooled, despite the noticeably different microstructure (Table 6), the heating time does not have a significant effects on the differences in the value of residual tensile strength—in both cases it remains on an almost identical level. In the case of a longer heating time, recrystallization of the defective ferritic matrix was observed, which translated into a slightly greater decrease in the UTS value.

Tempering at the temperature of 800 °C combined with subsequent water-cooling (WC/800/60 and WC/800/240 samples) did not result in the martensite → ferrite + pearlite phase transformation, despite reaching the heating temperature above A_1. As can be seen in the photos of the microstructure presented in Table 7, the material in this case still shows a typical martensite structure similar to the original one in the IS state. The cooling method applied, characterised by a high speed of thermal energy reception, had a significant impact on inhibition of the phase transformation.

Table 7. Microstructure of steel after heating at 800 °C for 60 min and 240 min with water cooling.

Sample Label	Image at a Distance of 3 mm from the Sample Edge	Image at a Distance of 7 mm from the Sample Edge
WC/800/60	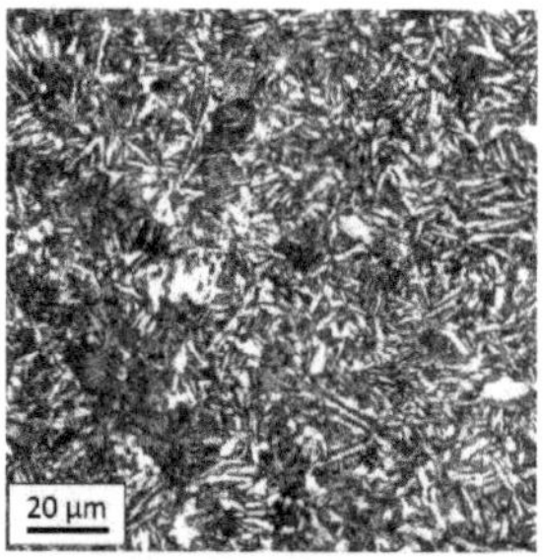	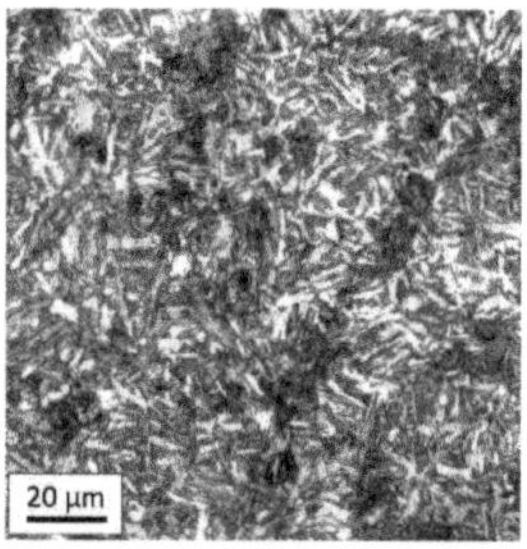
WC/800/240	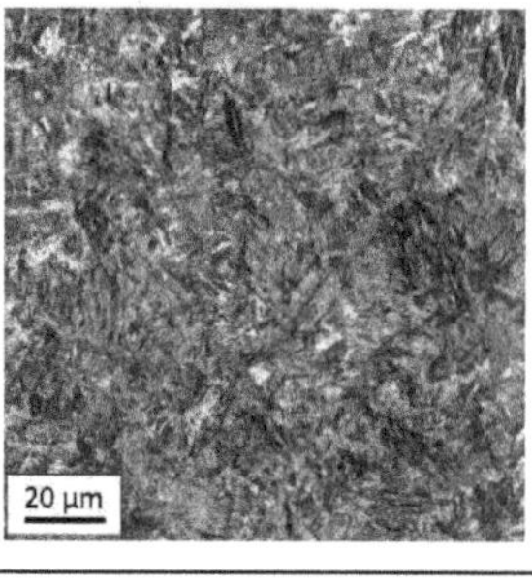	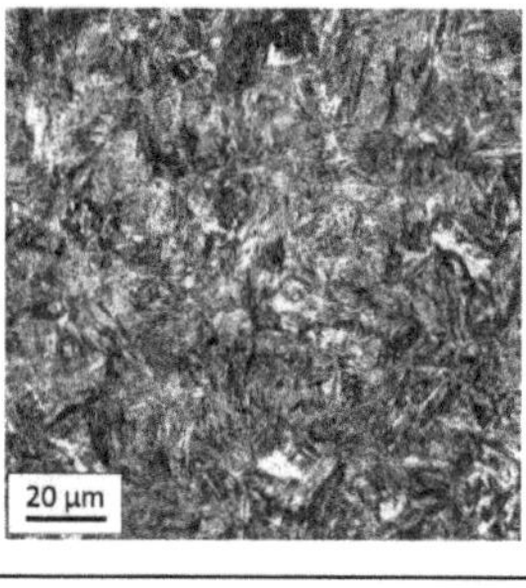

Rapid water cooling of the steel heated up to austenitisation temperature made it harden again. The degree of hardening turned out to be clearly dependent on the heating time. The martensite transformation did not fully take place within 60 min (residual austenite is still visible), which resulted in the value of UTS = 1064 MPa (maximum tensile force F_m = 261 kN—higher than the reference value for the sample in the initial state), with a simultaneous significant increase in hardness to HV = 361. Heating for 240 min resulted in austenitisation of the whole microstructure and the martensite transformation in the entire volume of the material, which translated into the value of UTS = 1064 MPa (maximum tensile force F_m = 261 kN—analogous as in the case of the heating time of 60 min), with another significant increase in hardness to the level of HV = 543.

This stage of research showed that, in the case of shock water-cooled samples, the duration of the heating time had a significant effect on increased hardness of the material and the use of rapid cooling noticeably influenced the value of residual tensile strength, which in this case exceeded the reference value obtained for bolts in the initial state IS (UTS = 1001 MPa).

3.1.3. Microstructural Analysis of Samples Heated at 1000 °C

Tempering at 1000 °C with air cooling (AC/1000/60 and AC/1000/240 samples, Table 8), similarly to the heat treatment at 800 °C, caused the martensite ➔ austenite phase transformation during the heating process and then the release of ferrite and pearlite within the boundaries of austenite grains during the phase of slow and free cooling.

Table 8. Microstructure of steel after heating at 1000 °C for 60 min and 240 min with air-cooling.

Sample Label	Image at a Distance of 3 mm from the Sample Edge	Image at a Distance of 7 mm from the Sample Edge
AC/1000/60	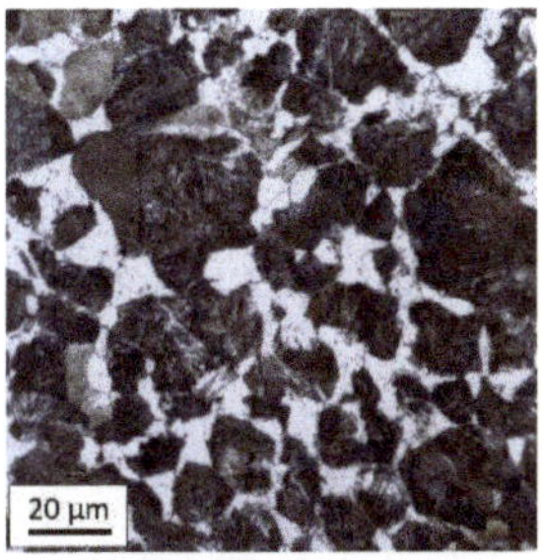	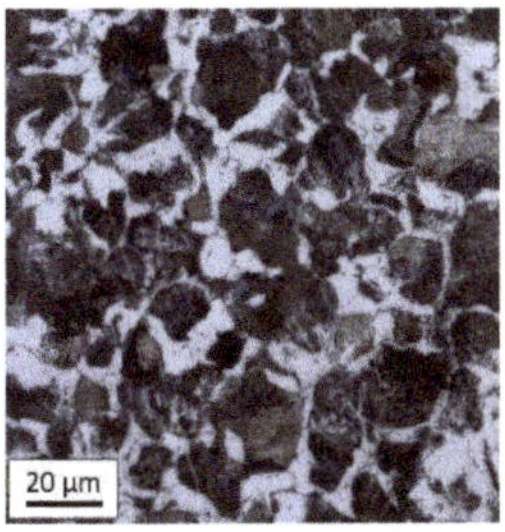
AC/1000/240	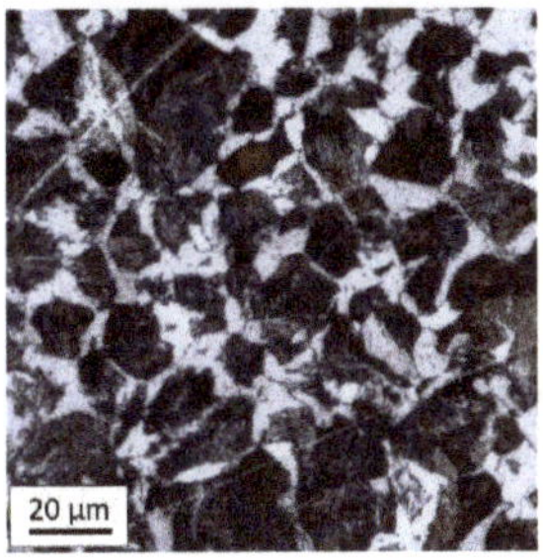	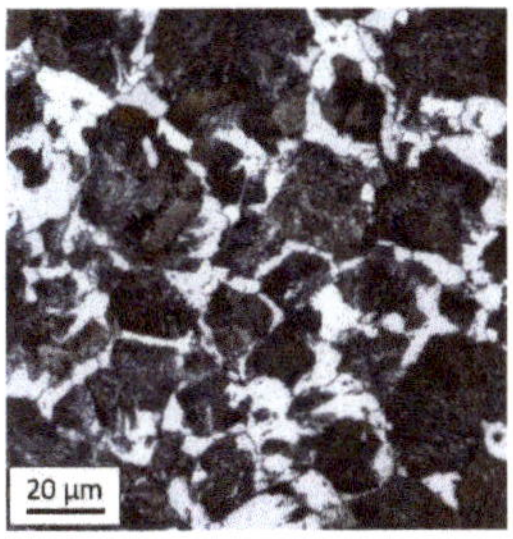

The share of ferrite (light areas) decreased noticeably, whereas the share of pearlite (dark areas) increased compared to samples heated at 800 °C and air-cooled (AC/800/60 and AC/800/240). A significant grain growth and recrystallization of the defective ferritic matrix were observed, which in turn resulted in a decrease in the residual ultimate tensile strength to the level of UTS = 619 MPa (maximum tensile force F_m = 152 kN), with HV = 204 for the heating time of 60 min and UTS = 576 MPa (maximum tensile force F_m = 142 kN) with a noticeably lower HV = 153 for the heating time of 240 min, respectively. Attention should be paid to a visible difference in hardness depending on the applied bolt heating time. In the case of samples cooling down naturally, an increased heating time results in a reduced value of residual tensile strength and hardness of the material. The same trend was also observed in the case of the AC/600 and AC/800 series samples, which may confirm that this trend is of a structured nature.

Heat treatment at the temperature of 1000 °C with water-cooling (WC/1000/60 and WC/1000/240 samples), just as in the case of heat treatment at 800 °C, did not lead to the martensite ➔ ferrite + pearlite phase transformation, although a temperature considerably higher than A_1 was reached. The high cooling rate made the steel harden again such that the material still had the martensite structure similar to the initial one IS (Table 9).

Table 9. Microstructure of steel after heating at 1000 °C for 60 min and 240 min with water-cooling.

Sample Label	Image at a Distance of 3 mm from the Sample Edge	Image at a Distance of 7 mm from the Sample Edge
WC/1000/60	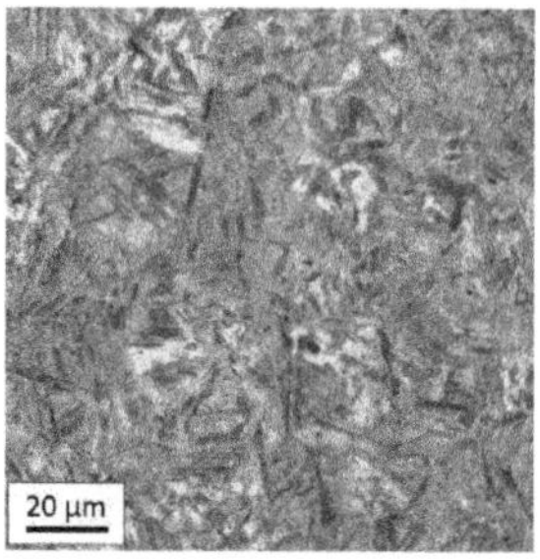	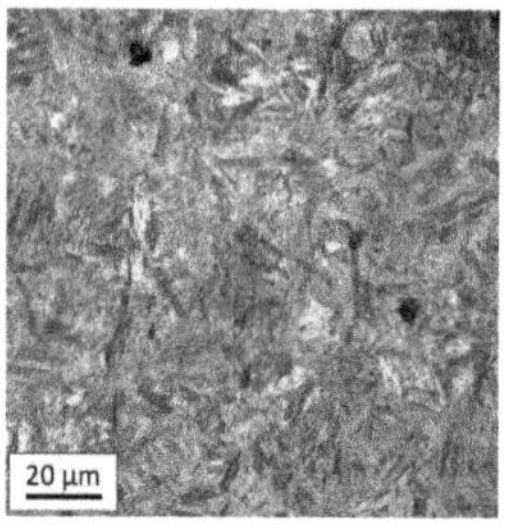
WC/1000/240	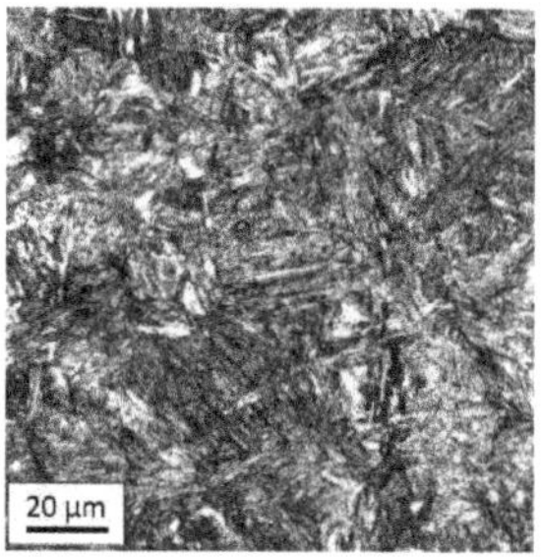	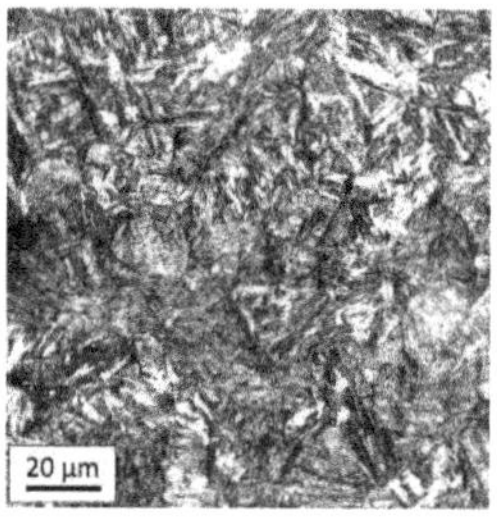

The high temperature of heat treatment contributed to the formation of coarse-grained martensite. The degree of hardening turned out to be significantly dependent on the heating time. Within 60 min, complete austenitisation took place and coarse-grained martensite was obtained in the entire volume, which resulted in a significant increase in the residual ultimate tensile strength value to the level of UTS = 1178 MPa (maximum tensile force F_m = 289 kN) and exceeded the reference value for the sample in the initial state with HV = 540. Heating for 240 min turned out to be too long and caused the grains to grow, which translated into a decrease in the UTS value to the level of 869 MPa (maximum tensile force F_m = 213 kN) with a simultaneous considerable decrease in hardness to HV = 210.

3.2. *Testing with the Use of X-ray Diffraction (XRD)*

Steel of type 32CrB3 contains 0.74% of chromium in its chemical composition. Chromium increases the temperature of the austenite transformation, slightly decreases ferrite hardness, increases the amount of residual austenite in the quenching process, and increases impact strength [19]. Heat treatment of steel, in which chromium is the alloying element, may contribute to the formation of carbides, which in the end has a noticeable effect on mechanical properties of steel. Carbides formed in steel are a hard and brittle phase. They are formed as a result of the solubility level of carbon in austenite and ferrite changing along with the temperature change. The presence of carbides in steel most often increases its hardness, yield point, and tensile strength. The carbides are also responsible for the secondary hardness effect, i.e., an increase in steel hardness during tempering. At the same time, their presence can have an adverse effect on impact resistance, ductility, and fracture resistance of steel [19]. Increased hardness and residual tensile strength in the case of the majority of samples heated at the temperature of 800 °C and 1000 °C might lead to a presumption that the presence of carbides in the material structure could be responsible for some of these changes. In order to exclude the presence of carbides in the steel structure before and after the heating process at the temperature of 1000 °C, XRD tests were carried out. The tests were carried out on both samples cooled naturally in air (AC) and those shock-cooled in water (WC). An analysis of the XRD results showed that obtained diffractograms had not differed in the number of peaks, but only in their intensity, which confirmed the fact that there were no carbides present in the steel structure before and after the heating process for the case of air-cooled and water-cooled samples. Considering the size of this article, the detailed results of XRD tests are not presented in the wider range.

3.3. *Quantitative Analysis of the Microstructure*

In respect to each of the phases shown in the pictures of the bolt microstructure, which were taken with the use of a digital light microscope (at the magnification of ×100) on the samples specified in Table 1, a quantitative analysis was carried out in order to determine the number of grains visible in the photo, their density, mean grain area, mean, and equivalent grain diameter. The analysis was performed separately for the ferrite and pearlite phases. The tests were aimed at confirming the previously observed qualitative change in the grain size in quantitative terms and aimed at linking these changes with changes in residual mechanical properties of the analysed samples. Due to limited funds, the analysis was performed only in respect to selected samples.

Collective results showing the dependence of microstructure indices, i.e., mean diameters of ferrite and pearlite grains, HV hardness, and tensile strength, on heat treatment parameters are summarised in Table 10 and Figure 1.

Table 10. Results of measurements of the mean equivalent diameter of ferrite and pearlite grains, hardness, and strength of respective series of samples.

Sample Label	Equivalent Diameter [μm]		Hardness	UTS
	Ferrite	Pearlite	[HV]	[MPa]
AC/800/60_3 mm	2.67	4.01	200.41	594.53
AC/800/60_7 mm	3.19	6.20	200.41	594.53
AC/800/240_3 mm	8.85	11.50	196.46	585.47
AC/800/240_7 mm	7.55	11.30	196.46	585.47
AC/1000/240_3 mm	6.19	13.70	152.98	575.88
AC/1000/240_7 mm	6.62	13.00	152.98	575.88
WC/800/60_3 mm	1.77	not measured	361.26	1063.90
WC/800/60_7 mm	3.39	not measured	361.26	1063.90

The measurements showed that in the case of air-cooled samples, the size of the pearlite grains increased along with an increase in the temperature level as well as the time of thermal exposure. However the grain growth was not uniform across the entire width of the bolt shank. For air-cooled samples, both in the case of ferrite and pearlite grains, the extension of the heating time at 800 °C from 60 min to 240 min resulted in an over threefold increase in the grain equivalent diameter. Prolonged annealing at the temperature of 1000 °C resulted in a further growth of pearlite grains to approximately 13.0 μm in diameter, with a simultaneous approximately 20% decrease in the ferrite grain size.

In the case of water-cooled samples, the ferrite grain size is clearly smaller than in the case of air-cooled samples corresponding to them in terms of heat treatment conditions. This confirms the previous observations made on the basis of the visual analysis of the pictures included in Table 9. The water shock cooling of the samples prevented the phase transformation of austenite into pearlite and inhibited the growth of ferrite grains. The impact of the shock cooling on the size of ferrite grains was noticeable—near the outer walls of the sample shank, the grain diameter is almost half the size of those near the bolt axis. In the case of air-cooled samples, along with an increase in temperature and heat-exposure time, a slight downward trend in the value of the residual tensile strength and the hardness of the bolt material was also clearly visible. The water shock cooling re-hardened the bolts and was followed by a sharp increase in both the UTS and HV values (Figure 1).

3.4. Analysis of the Correlation between Hardness and Residual Tensile Strength

The purpose of the analysis was to investigate veracity of the linear correlation described by the Formula (1) between tensile strength UTS and hardness HV of steel subjected to heat treatment during the production process and then subjected to secondary thermal treatment, e.g., as a result of exposure to thermal effects of a fire. The relationship (1) has been confirmed so far by numerous tests carried out almost exclusively on samples of commonly used structural steels working in normal thermal conditions. The available literature does not provide any information proving its correctness in relation to the value of residual tensile strength characteristic of the material of high-strength steel bolts after the fire exposure.

In order to better illustrate the effect of secondary heat treatment parameters on the relationship between hardness HV and residual tensile strength of the bolt steel, a comprehensive diagram has been presented in Figure 2. It shows a clear linear relationship between the values of hardness and tensile strength but it meets the criterion described by the scaling factor $3.2 \div 3.5$ that is not in the entire domain of determinacy. In the case of air-cooled samples, the value of this factor obtained in the tests fluctuates in the range of $3.0 \div 3.8$ and in the case of water-cooled samples the factor fluctuates in the range of $2.0 \div 4.1$, respectively. In the case of air-cooled samples, attention should be paid to a noticeable trend of a decrease in strength (in relation to the reference value, characteristic of the material in its initial state) accompanying a temperature and heating time increase. The hardness parameter also shows a similar trend. In the case of shock water-cooled samples, this trend which is characteristic of air-cooled samples, is reversed. Along with an increase in the temperature and heating time, the values of the residual tensile strength and hardness of the bolt steels increase. Some regularity of this trend is locally disturbed only in the case of the WC/600/240 and WC/1000/240 samples, which may, however, be caused by a systematic error due to the small size of the sample. Confirmation of this thesis would nevertheless require further research. In the case of shock water-cooling of samples (simulation of a rescue and firefighting operation), both residual tensile strength and hardness of the bolt material increases significantly and under certain conditions even exceeds the reference values characteristic of the tested high-strength bolts in their initial state. On the one hand, it can be perceived as a desirable effect due to the significant strengthening of the material and, on the other hand, it can be perceived as a negative effect due to an increase in its brittleness, which translates into the possibility of an abrupt form of failure in the event of overloading the bolts in the joint during an operation phase after a fire.

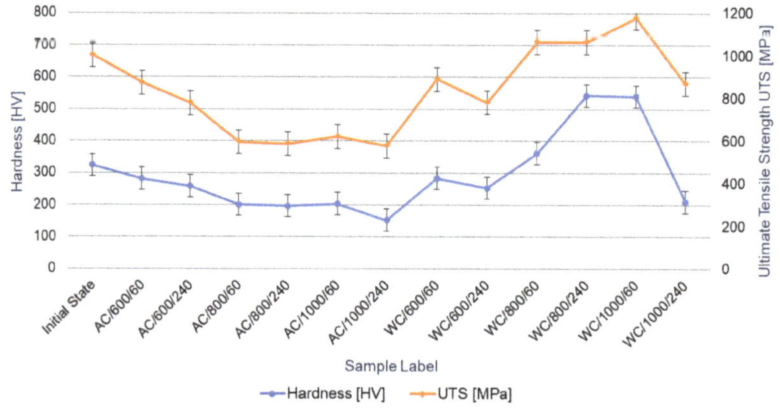

Figure 2. Diagram of correlations between hardness HV and residual tensile strength of the bolt steel after secondary heat treatment depending on parameters of this treatment.

In order to determine the effect of the cooling method, temperature level and heating time on the disturbance of the relationship between the hardness of the bolt material and its residual tensile strength and the relationships between these two values in different configurations are shown in Figures 3 and 4.

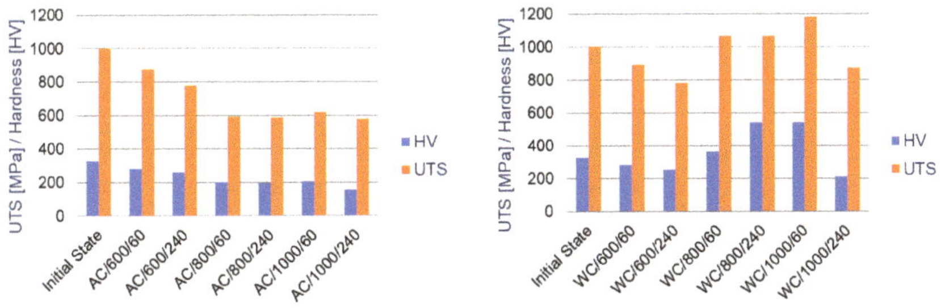

(**a**) (**b**)

Figure 3. Diagram of dependencies between bolt steel hardness and bolt residual tensile strength on heating time for samples: (**a**) naturally air-cooled; (**b**) shock water-cooled.

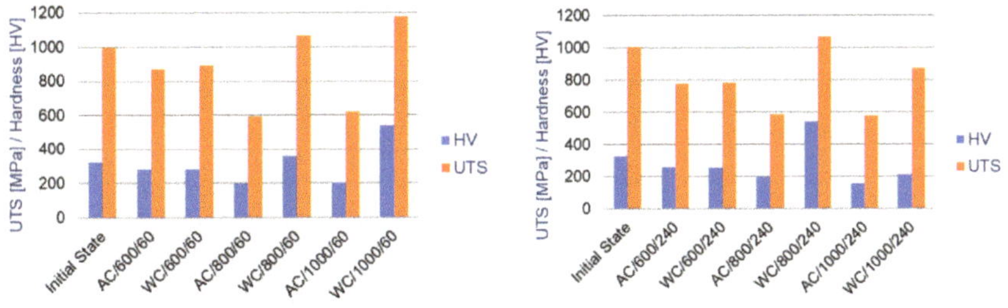

(**a**) (**b**)

Figure 4. Diagram of dependencies between bolt steel hardness and bolt residual tensile strength on heating time for samples subjected to secondary heat treatment for the time of: (**a**) 60 min; (**b**) 240 min.

When analysing the diagrams shown in Figure 3, it can be observed that a negative effect of long heating of a particular series of samples is clearly visible—both in the value of residual ultimate tensile strength UTS and hardness HV. The same trend showing the effect of the heating time on stability of the analysed material properties is visible for each temperature level. Excessively long heating results in a decrease in residual strength. As a consequence, material hardness is also reduced.

The analysis of Figure 4 shows that in the temperature range up to 600 °C, both in the case of air-cooled bolts and those water-cooled, no effect of the heating time on stability of the analysed mechanical properties of the samples has been identified. Corresponding pairs of graphs for the AC/600/60 and WC/600/60 and AC/600/240 and WC/600/240 samples do not differ from each other. The only difference between the bars of the graphs shown in Figure 3a,b is in the UTS and HV values. Even the UTS/HV inter-relationships for the corresponding pairs of samples are almost identical. This confirms the previous observations made on the basis of the analysis of microstructure pictures, described in Section 3.1.1 of this article.

An increase in the heating temperature to 800 °C and higher makes the differences between the air-cooled and water-cooled samples clearly visible. Under similar thermal conditions, water shock cooling results in achieving a much higher residual tensile strength of the sample material—UTS and hardness HV compared to the sample freely cooling in the air. This results from the fact that the samples are re-hardened and the processes of phase transformations are stopped.

4. Discussion

A comprehensive approach to the analysis of the effect of simulated natural fire conditions on microstructural changes in the material of construction bolts and also on their key strength properties, taking into account at the same time the effect of thermal conditions, exposure time, and cooling method, is presented in this paper and cannot be found in available literature. In particular, with regard to the analysis of the influence of exposure time on the given thermal conditions, the conducted research is somewhat unique. In this context, this work brings a completely new value to the state of knowledge in this area and may really contribute to the progress of work on methods of post-fire assessment of quality and reliability of structures.

The study confirmed that, as a result of fire exposure, the microstructural and mechanical characteristics of high-strength bolts undergo significant changes. These changes can have a negative impact on the safety and reliability of steel structures, as well as the manner they behave after a fire.

The bolts heated to a temperature exceeding 600 °C and cooled naturally in the air, as a result of fire exposure undergo a martensite ➔ austenite phase transformation during the heating process and then ferrite and pearlite are released within the boundaries of austenite grains in their structure. This structural modification renders steel softer and bolt load-bearing capacity considerably lower. However, as a rule and at the expense of increased plasticity, an overloaded bolt does not rapidly fail. Bolts that underwent such a heating cycle permanently lose their original strength properties. This, in general, confirms the observations made by Kirby [1,2], Kodur et al. [5,6], and Yahyai [7]. However, each of them in their considerations ignored the effect of rapid cooling, which is a natural consequence of the fire-fighting action, as well as the related consequences concerning both the microstructure and mechanical properties of the tested bolts. Kodur et al. [5,6], similar to Yahyai et al. [7] did not focus on detailed microstructural studies, and they only tried to link the bolt failure model with the macroscopic image of the fracture surface. In the case of screws naturally cooled in the air after the annealing process at a temperature exceeding 600 °C, similar to observations of Sajid et al. [10], the significant growth of pearlite and ferrite grains was observed leading to a reduction in the ultimate tensile strength. This clearly confirms the truth of the Hall–Petch formula (Equation (2)) [18] not only in relation to the yield point of steel but also to the value of the UTS.

Bolts heated to a temperature exceeding 600 °C and shock cooled with water (e.g., during firefighting operation in a real fire) retain a martensite structure, similar to the original one, because they are re-hardened through rapid cooling. Due to the sudden reception of thermal energy, they do not undergo the microstructural change characteristic of the martensite ➔ ferrite + pearlite phase transformation, despite reaching a temperature significantly higher than A_1. The steel of the bolt hardens and the temporary load capacity of bolts can, as a result of the re-hardening, even exceed the load capacity that the bolt had in its initial state or be very close to this value. This is performed at the expense of increased brittleness of the bolt such that, in the event of overload, the bolt can be expected to fail rapidly. Although the conducted research did not carry out such detailed quantitative analysis as performed by Chi and Peng [9], the obtained results confirmed the essence of their observations with regard to the batch of water-cooled bolts.

The conducted research confirmed the possibility of using microstructural tests for a post-fire assessment of steel structures and an attempt to reconstruct fire scenarios. However, referring to the concept of the use of metallurgical changes occurring in the screw material for the diagnosis of buildings damaged by fire proposed, inter alia, by Kirby [1,2], Chi and Peng [9], and Sajid et al. [10], it should be stated emphatically that unless the precise fire temperature is known, an assessment based solely on microstructural tests may not be sufficient or may be flawed with significant errors. This may be the case, for example, when the image of the bolt material microstructure indicates a martensite structure. In the case of high-strength bolts, this indication is typical for bolts in their initial state, bolts heated to a temperature lower than 600 °C, and bolts heated to a much higher temperature and rapidly cooled with water. In this case, a reliable assessment cannot be made without conducting additional destructive strength testing. This example shows that the post-fire assessment of bolts should be performed in an extremely reasonable and careful method.

The strength tests of the initial phase (which for the sake of clarity of the article content have not been included here) confirmed the earlier observations made, among others, by Kirby [1,2] and Kodur et al. [5,6] that the residual mechanical properties of bolts subjected to fire, due to their heat treatment history, differ from those determined for carbon structural steels given in the standards for designing structures [20,21]. The resistance reduction factors given in Annex D of [20] applicable during connection design were determined from the fire tests of bolts conducted for British Steel [1,2] by Kirby. Although the reduction factors deduced from the tensile and shear tests were different, in order to simplify the design guidance the conservative results were applied for both types of connections—those loaded in tension and those loaded in shear. Theoretically, applying these guidelines to bolts can lead to conservative estimates but one has to realize that these provisions were based on test results of air-cooled bolts only. It cannot be forgotten that the increased brittleness of bolts and elements arising from the fire-fighting action increases the risk of an abrupt and progressive collapse caused by the inability of the structure to redistribute internal forces by creating local yielding areas. This, in turn, may result in an increase in the threat to lives and the health of users of the building, who are the potential victims of a fire and members of rescue teams. In the case of new construction design, the use of Annex D [20] in connection with design likely leads to safer estimates. The observation of existing objects destroyed by a fire but designed on the basis of earlier standards leads to completely opposite conclusions than discussed above. It is the connections in older buildings that seem to be the weakest link in the entire structure. The authors draw attention to the importance of this problem, especially in the case of facilities such as shopping centres, galleries, arcades, or other public facilities made of steel structures. However, it should also be noted that the quality of steel products has improved since Kirby's publication of the results [1,2] and the reduction factors may need to be updated based on the results of up-to-date tests.

5. Limitations, Recommendations, and Suggestions for Further Research

The outcomes show that the method of carrying out a rescue and firefighting operation may be of key importance for the safety of structure and people staying in a building in fire. In the case of structures subject to dynamic loads, water cooling of bolts should be avoided as this will render them more susceptible to brittle fracture. In the case of fire-damaged structures that have already undergone significant deformations, potential gains and losses should be assessed on a somewhat real-time basis. If members have not been deformed by a fire and there is a real chance of renovating and re-using, then the risk of cooling the bolts may be taken in order to increase their load capacity in real time. Of course, during the reconstruction or renovation of the structure, such bolts must be replaced with new ones with predictable strength properties. If the structure cannot be saved, its stability is compromised or it has undergone significant deformations preventing it from being reused, bolts should not be hardened to permit their slow plastic failure at the moment of being overloaded.

A big challenge in the logistic, economic, and scientific sense is the study of complete joints and connections performed in their natural scale. The studies of butt joints [22–24] and lap friction joints [25] carried out so far and described in the literature focus mostly on establishing the model of failure and analysis of the behaviour of these joints under load. They also form the basis for the validation and verification of numerical models that try to map the physical behaviour of a connection or joint in a real structure. These studies do not take into account the drop in pre-stressing force as a result of fire in connections that were pre-stressed at the assembly stage. This gap is worth paying more attention to in terms of conducting completely new research. It is also worth considering, in the future, a wide range of tests utilizing stainless steel and fire-resistant bolts. The heat-resistant and creep-resistant steels are able to withstand temperatures up to approximately 1150 °C for a long period of time. They obtain their heat resistance thanks to a wide range of alloy additives: aluminium, chromium, silicon, molybdenum, vanadium, tungsten, titanium, and cobalt and increases the energy of interatomic bonds. In particular, unlike conventional carbon steels, austenitic alloys used for creep-resistant steels are characterised by extremely low susceptibility to structural changes in long-term operation at high temperatures. The differences in resistance to high temperatures between carbon steels and fire-resistant steels, in the context of their susceptibility to microstructural changes, are best illustrated by comparing the diagrams of their specific heat as a function of temperature. The characteristic peak at 735 °C in the case of carbon steels, resulting from their metallurgical changes, does not occur in the diagram for fire-resistant steels. Although the use of heat-resistant and creep-resistant steels in modern construction is currently rather niche, in just a few years, due to technical progress, the development of engineering of structural materials or technological methods for modifying the microstructure of construction materials [26,27] could become quite common.

6. Conclusions

The presented research is of a practical nature with great application potential for engineering practice.

The results obtained can be helpful for the purpose of assessing structures that have survived a fire without any major damage, in the context of the possibility of reusing selected elements, and those that have been damaged by fire. In the latter case, the results can be used to recreate a fire development scenario and to estimate maximum values of fire temperatures in a situation where they have not been measured by firefighting and rescue services.

The microstructural tests confirmed that under environmental conditions corresponding to a simulated fire situations and an accompanying firefighting operation, significant structural changes occur in the material of bolts that are strongly dependent on the temperature reached, the time of exposure to fire conditions, and the method of cooling. These changes result in modifications of residual strength properties and are crucial from the point of view of structural safety, in particular tensile strength and correlated hardness. Microstructural changes significantly affect how the bolt material behaves, which usually determines how the structure will fail in the event of a potential collapse.

The undertaken research and obtained results indicate that this work should be continued with focus on development of detailed guidelines for designing bolted joints while simultaneously taking into account the effects of fire.

Author Contributions: Conceptualization, P.A.K.; methodology, P.A.K. and M.W.; formal analysis, P.A.K.; investigation, P.A.K. and M.W.; resources, P.A.K. and M.W.; data curation, P.A.K. and M.W.; writing—original draft preparation, P.A.K. and M.W.; writing—review and editing, P.A.K.; visualization, P.A.K. and M.W.; supervision, P.A.K.; project administration, P.A.K.; funding acquisition, P.A.K. All authors have read and agreed to the published version of the manuscript.

Funding: This research received no external funding.

Institutional Review Board Statement: Not applicable.

Informed Consent Statement: Not applicable.

Data Availability Statement: Data sharing not applicable.

Acknowledgments: We would like to thank the ASMET Sp. z o.o. company seated in Reguły near Warsaw and, in particular, Andrzej Sajnaga—the Owner and President of the Management Board—and Andrzej Czupryński—the Technical Director. Without considerable logistic and technical support from the Company, which provided the bolts for the research, the experimental program carried out with the aim of testing bolts exposed to fire conditions would not have been possible.

Conflicts of Interest: The authors declare no conflict of interest. The funders had no role in the design of the study; in the collection, analyses, or interpretation of data; in the writing of the manuscript, or in the decision to publish the results.

References

1. Kirby, B.R. The behaviour of high-strength grade 8.8 bolts in fire. In *Technical Report SL/HED/R/S1792/1/92/D*; British Steel Technical Swinden Laboratories: Rotherham, UK, 1992.
2. Kirby, B.R. The behavior of high-strength grade 8.8 bolts in fire. *J. Constr. Steel Res.* **1995**, *33*, 3–38. [CrossRef]
3. González, F.; Lange, J. Behaviour of high strength grade 10.9 bolts under fire conditions. In *Proceedings of the International Conference on Application of Structural Fire Design, Prague, Czech Republic, 19–20 February 2009*; Pražská Technika; Czech Technical University in Prague: Prague, Czech Republic, 2009; pp. 392–397.
4. Hanus, F.; Zilli, G.; Franssen, J.-M. Behaviour of grade 8.8 bolts under natural fire conditions—Tests and model. *J. Constr. Steel Res.* **2011**, *67*, 1292–1298. [CrossRef]
5. Kodur, V.; Kand, S.; Khaliq, W. Effect of temperature on thermal and mechanical properties of steel bolts. *J. Mater. Civ. Eng.* **2012**, *24*, 765–774. [CrossRef]
6. Kodur, V.; Yahyai, M.; Rezaeian, A.; Eslami, M.; Poormohamadi, A. Residual mechanical properties of high strength steel bolts subjected to heating-cooling cycle. *J. Constr. Steel Res.* **2017**, *131*, 122–131. [CrossRef]
7. Yahyai, M.; Kodur, V.; Rezaeian, A. Residual mechanical properties of high strength steel bolts after exposure to elevated temperature. *J. Mater. Civ. Eng.* **2018**, *30*. [CrossRef]
8. Satheeskumar, N.; Davison, J.B. Robustness of steel joints with stainless steel bolts in fire. *Int. J. Adv. Struct. Eng.* **2014**, *6*, 161–168. [CrossRef]
9. Chi, J.-H.; Peng, P.-C. Using the microstructure and mechanical behavior of steel materials to develop a new fire investigation technology. *Fire Mater.* **2017**, *41*, 864–870. [CrossRef]
10. Sajid, H.U.; Kiran, R.; Naik, D. Microstructure-mechanical property relationships for post-fire structural steels. *J. Mater. Civ. Eng.* **2020**, *32*. [CrossRef]
11. Haiko, O.; Kaijalainen, A.; Pallasburo, S.; Hannula, J.; Porter, D.; Liimatainen, T.; Kömi, J. The effect of tempering on the microstructure and mechanical properties of a novel 0.4C press-hardening steel. *Appl. Sci.* **2019**, *9*, 4231. [CrossRef]
12. Xie, Z.; Song, Z.; Chen, K.; Jiang, M.; Tao, Y.; Wang, X.; Shang, C. Study of nanometer-sized precipitation and properties of fire resistant hot-rolled steel. *Metals* **2019**, *9*, 1230. [CrossRef]
13. Regulation of the Minister of Infrastructure of 12 April 2002 on the technical requirements to be met by buildings and their location. *J. Laws Repub. Pol.* **2019**, 1065. Available online: https://www.global-regulation.com/translation/poland/3353940/regulation-of-the-minister-of-infrastructure-of-12-april-2002-on-technical-conditions%252c-which-should-correspond-to-the-buildings-and-their-location.html (accessed on 15 April 2021). (In Polish)
14. Construction Law Act of 7 July 1994. *J. Laws Repub. Pol.* **2020**, 1333. Available online: https://www.buildup.eu/en/practices/publications/polish-construction-law (accessed on 15 April 2021). (In Polish)
15. Regulation (EU) No 305/2011 of the European Parliament and of the Council of 9 March 2011 Laying Down Harmonised Conditions for the Marketing of Construction Products and Repealing Council Directive 89/106/EEC (Text with EEA Relevance), Official Journal of the European Union L 88/5. Available online: http://data.europa.eu/eli/reg/2011/305/oj (accessed on 15 April 2021).
16. Inspection Certificate 3.1, acc. to EN 10204:2004, Issued for Wire Rod 32CrB3 with Diameter of 21 mm, by CMC Poland Sp. z o.o., 42–400 Zawiercie, ul. Piłsudkiego 82.
17. Bolton, W.; Higgins, R.A. *Materials for Engineers and Technicians*, 7th ed.; Routledge Taylor & Francis Ltd.: Abingdon-on-Thames, UK, 2020; ISBN 978-0367535506.
18. Bhadeshia, H.; Honeycombe, R. *Steels: Microstructure and Properties*, 3rd ed.; Butterworth-Heinemann: Oxford, UK, 2006; ISBN 9780080462929.
19. Meyers, M.A.; Chawla, K.K. *Mechanical Behavior of Materials*, 2nd ed.; Cambridge University Press: Cambridge, UK; ISBN 9780521866750.
20. *EN 1993-1-2 Eurocode 3: Design of Steel Structures–Pat 1–2: General Rules–Structural Fire Design*; European Committee for Standardization: Brussels, Belgium, 2005.
21. *EN 1994-1-2 Eurocode 4: Design of Composite Steel and Concrete Structures–Part 1–2: General Rules–Structural Fire Design*; European Committee for Standardization: Brussels, Belgium, 2005.
22. Sarigoglu, M. Experimental evaluation of the post-fire behaviour of steel T-component in the beam-to-column connection. *Fire Saf. J.* **2018**, *969*, 153–164. [CrossRef]
23. Quiang, X.; Jiang, X.; Bijlaard, F.S.K.; Kolstein, H.; Luo, Y. Post-fire behaviour of high strength steel endplate connections—Part.1: Experimental study. *J. Constr. Steel Res.* **2015**, *108*, 82–93. [CrossRef]
24. Al-Jabri, K.S.; Davison, J.B.; Burgess, I.W. Performance of beam-to-column joint in fire—A review. *Fire Saf. J.* **2008**, *43*, 50–62. [CrossRef]

25. Liu, H.; Liu, D.; Chen, Z.; Yujie, Y. Post-fire residual slip resistance and shear capacity of high-strength bolted connection. *J. Constr. Steel Res.* **2017**, *138*, 65–71. [CrossRef]
26. Maślak, M.; Skiba, R. Fire resistance increase of structural steel through the modification of its chemical composition. *Procedia Eng.* **2015**, *108*, 277–284. [CrossRef]
27. Seo, J.E.; Cho, L.; Estrin, Y.; De Cooman, B.C. Microstructure-mechanical properties relationship for quenching and partitioning (Q&P) processed steel. *Acta Mater.* **2016**, *113*, 124–139. [CrossRef]

materials

MDPI

Article

Effect of Boron and Vanadium Addition on Friction-Wear Properties of the Coating AlCrN for Special Applications

Huu Chien Nguyen [1], Zdeněk Joska [1,*], Zdeněk Pokorný [1], Zbyněk Studený [1], Josef Sedlák [2], Josef Majerík [3], Emil Svoboda [1], David Dobrocký [1], Jiří Procházka [1] and Quang Dung Tran [4]

[1] Department of Mechanical Engineering, Faculty of Military Technology, University of Defence, 612 00 Brno, Czech Republic; huuchien.nguyen@unob.cz (H.C.N.); zdenek.pokorny@unob.cz (Z.P.); zbynek.studeny@unob.cz (Z.S.); emil.svoboda@unob.cz (E.S.); david.dobrocky@unob.cz (D.D.); jiri.prochazka@unob.cz (J.P.)

[2] Department of Industrial Engineering and Information Systems, Faculty of Management and Economics, Tomas Bata University in Zlin, 760 01 Zlin, Czech Republic; sedlak@utb.cz

[3] Faculty of Special Technology, Alexander Dubcek University of Trencin, 91101 Trencin, Slovakia; jozef.majerik@tnuni.sk

[4] Faculty of Mechanical Engineering, Le Quy Don Technical University, Hanoi 100000, Vietnam; tranquangdung79@lqdtu.edu.vn

* Correspondence: zdenek.joska@unob.cz; Tel.: +420-973-442-989

Citation: Nguyen, H.C.; Joska, Z.; Pokorný, Z.; Studený, Z.; Sedlák, J.; Majerík, J.; Svoboda, E.; Dobrocký, D.; Procházka, J.; Tran, Q.D. Effect of Boron and Vanadium Addition on Friction-Wear Properties of the Coating AlCrN for Special Applications. *Materials* **2021**, *14*, 4651. https://doi.org/10.3390/ma14164651

Academic Editors: Marcin Wachowski, Henryk Paul and Sebastian Mróz

Received: 14 July 2021
Accepted: 13 August 2021
Published: 18 August 2021

Publisher's Note: MDPI stays neutral with regard to jurisdictional claims in published maps and institutional affiliations.

Abstract: Cutting tools have long been coated with an AlCrN hard coating system that has good mechanical and tribological qualities. Boron (B) and vanadium (V) additions to AlCrN coatings were studied for their mechanical and tribological properties. Cathodic multi-arc evaporation was used to successfully manufacture the AlCrBN and AlCrVN coatings. These multicomponent coatings were applied to the untreated and plasma-nitrided surfaces of HS6-5-2 and H13 steels, respectively. Nanoindentation and Vickers micro-hardness tests were used to assess the mechanical properties of the materials. Ball-on-flat wear tests with WC-Co balls as counterparts were used to assess the friction-wear capabilities. Nanoindentation tests demonstrated that AlCrBN coating has a higher hardness (HIT 40.9 GPa) than AlCrVN coating (39.3 GPa). Steels' wear resistance was significantly increased by a hybrid treatment that included plasma nitriding and hard coatings. The wear volume was 3% better for the AlCrBN coating than for the AlCrVN coating on H13 nitrided steel, decreasing by 89% compared to the untreated material. For HS6-5-2 steel, the wear volume was almost the same for both coatings but decreased by 77% compared to the untreated material. Boron addition significantly improved the mechanical, tribological, and adhesive capabilities of the AlCrN coating.

Keywords: H13; HS6-5-2; AlCrBN; AlCrVN; nanohardness; friction; wear resistance; adhesion

1. Introduction

Today, great emphasis is paid to corporate environmental policies [1], with innovative usage of plasma nitriding and PVD coatings replacing traditional, less ecologically friendly techniques. Due to their high hardness, outstanding wear resistance, superior corrosion, oxidation resistance, and good thermal stability, AlCrN thin coatings have been more important in industry over the last few decades [2,3]. At high temperatures, AlCrN coatings exhibit exceptional oxidation resistance due to the production of protective mixed protective oxides Al_2O_3, Cr_2O_3 [3,4]. As a result, they are commonly employed in the automobile sector as abrasion-resistant layers (e.g., valves, tappets, and camshafts) or as protective coatings for forming and machining tools [4,5]. At room temperature, AlCrN coatings had a high friction coefficient of 0.7, which climbed to 1.0 at temperatures above 500 °C, a temperature range often encountered in cutting operations [6]. The addition of alloying elements such as Si, B, and V [2,7,8] could be one way to significantly improve the film characteristics. The components Si and B have been shown to enhance the creation of a nanocomposite structure in AlCrN coatings, improving hardness, toughness, and/or wear resistance [9]. Due to the creation of a V_2O_5-Magnéli oxide phase at high temperatures, the inclusion of V proved to be effective in reducing friction, especially at high temperatures [2].

Many studies have looked into AlCrVN coatings to increase the friction-wear properties of materials [2,10,11]. Because of the formation of V_2O_5 oxide during tribological testing at extreme temperatures, the AlCrVN coatings have a low friction coefficient (0.2–0.3) and good wear resistance at high temperatures (700 °C). The AlCrVN coating greatly increased cutting tool performance while also boosting anti-wear ability due to the high hardness and produced lubricating V_2O_5 coatings [11,12]. The findings indicate that AlCrVN hard coatings have the potential to improve the wear resistance of cutting tools, forming, mechanical, and weapon parts [10,11].

Sato et al. [7] and Nose et al. [13] reported that boron addition improved the mechanical properties of AlCrN coatings by combining the effects of solid solution hardening, grain size refinement (Hall–Petch hardening), and formation of nanocomposite structure, where a-BNx tissue phase embedded the AlCrN crystallites. The nanocomposite structure of AlCrBN coatings, in which nano-sized fcc AlCrN grains are surrounded by a thin BNx tissue phase, greatly increased the hardness of the AlCrN coatings while reducing compressive residual stress [14]. At room temperature and at the evaluated temperature (700 °C), AlCrBN coatings have recently been shown to have superhardness and high wear resistance [9]. Boron addition reduced the grain size of the AlCrN coatings from 40 nm to around 10 nm, and the internal stress of the coating system was lowered by more than 50% [15]. In comparison to AlCrN and AlCrTiN coatings, the AlCrBN coating resists gear hobbing against crater wear the best [15].

It has been observed that a hybrid surface treatment consisting of plasma nitriding and hard coatings significantly improved the mechanical and friction-wear properties of materials, as well as the coating adhesion strength [16–18]. Plasma nitriding, which generates a harder barrier between the soft substrate and the hard coating [19], plays a significant role in this hybrid surface treatment [16]. The nitrided hardened layer considerably reduces the plastic deformation [20] that happens underneath the coating. As a result, the coatings' durability and adhesion strength will be enhanced [17,21].

Although the microstructure, hardness, and wear resistance of AlCrBN and AlCrVN coatings have been extensively studied, the influence of surface treatments such as plasma nitriding and AlCrBN, AlCrVN coatings on the friction-wear properties of steel has still to be researched.

The purpose of this investigation was to see how a hybrid treatment consisting of plasma nitriding and AlCrVN, AlCrBN coating affected the friction-wear properties of H13 tool steel and HS6-5-2 high-speed tool steel. H13 hot-work steel is commonly used in both hot and cold work tooling. Cutting tools are commonly made of HS6-5-2 high-speed steel. Under the same conditions, AlCrVN and AlCrBN coatings were applied to the surfaces of both un-nitrided (only-coating) and nitrided (duplex coating) materials. The mechanical properties and wear resistance of the AlCrBN and AlCrVN multicomponent coatings were investigated and compared. The results from the un-nitrided materials, plasma-nitrided materials, and only-coated materials were compared to the results from the hybrid-surface-treated materials.

2. Materials and Methods

2.1. Materials

H13 steel and HS6-5-2 high-strength steel were chosen for use in the experiment. The chemical composition of the steel was assessed five times by Q4 TASMAN spark emission equipment (Bruker, Karlsruhe, Germany), with the averages used as data. The chemical compositions of H13 and HS6-5-2 steels are shown in Tables 1 and 2, respectively.

Table 1. H13 steel chemical composition (weight %).

	C	Mn	Si	Cr	Mo	V	P	S
ASTM A681	0.32–0.45	0.2–0.6	0.8–1.25	4.75–5.5	1.1–1.75	0.8–1.2	Max. 0.03	Max. 0.03
Measured	0.36	0.47	0.97	4.80	1.24	0.84	0.030	0.01

Table 2. HS6-5-2 steel chemical composition (weight %).

	C	Mn	Si	Cr	Mo	W	V	P	S
EN ISO 4957	0.8–0.88	Max. 0.4	Max. 0.45	3.8–4.5	4.7–5.2	5.9–6.7	1.7–2.1	Max. 0.03	Max. 0.03
Measured	0.82	0.35	0.23	4.50	5.35	1.94	7.00	0.028	0.010

Steel H13 samples have a diameter of 65 mm and a thickness of 6 mm, while steel HS6-5-2 samples have a diameter of 20 mm and a thickness of 3 mm. After heat treatment, all experimental samples were supplied. The heat-treated samples had a surface hardness of 52 HRC for steel H13 and 64 HRC for HS6-5-2 steel.

Surface roughness Ra of 0.11 μm for steel H13 and 0.04 μm for steel HS6-5-2 was obtained by grinding the samples using a Struers LaboSystem (Struers, Copenhagen, Denmark) grinder. Silicon carbide grinding paper grits 120, 220, 400, 600, and 800 were utilized for grinding. The cross sections of the specimens were polished to mirror surfaces using Leco CAMEO Disc Platinum (Leco Corporation, St. Joseph, MI, USA) 1, 2, 3, 4 and diamond polishing paste with grain size 1 μm by Leco PX-500 Grinder/Polisher (Leco Corporation, St. Joseph, MI, USA) for microstructure characterization. After that, a picric acid solution (1 g picric acid, 5 mL HCl acid in 100 mL ethanol) was used to etch the polished cross sections of steel HS6-5-2, and a 2 percent Nital etchant (a solution of ethanol and nitric acid) was used to etch the polished cross sections of steel H13. Plasma nitriding was carried out using PN60/60 RÜBIG equipment (Rubig GmBH, Wels, Austria) at 470 °C for 4 h with a mixture gas of $3H_2$:$1N_2$ at 280 Pa. UN and PN materials stand for un-nitrided and plasma-nitrided materials, respectively. The cathodic arc PVD process was utilized to coat AlCrBN and AlCrVN. The Pi411 equipment was used to coat AlCrBN and AlCrVN coatings on both un-nitrided and nitrided materials in the company LISS. Un-nitrided AlCrBN-coated and AlCrVN-coated materials are referred to as AlCrBN material and AlCrVN material, respectively. The PN/AlCrBN and PN/AlCrVN materials (respectively duplex-coated materials) are AlCrBN-coated and AlCrVN-coated nitrided materials.

2.2. Experiment Procedures

The surface roughness of materials was measured using absolute inductive position sensor by Talysurf CLI 1000 stylus profilometer (Taylor Hobson Ltd., Leicester, England). For evaluation of the surface, the parameter Ra was used—the arithmetic average of the absolute values of the profile heights over the evaluation length. The microstructures were observed on the cross sections of the specimens by Olympus DSX 500i opto-digital microscope(Olympus, Tokyo, Japan). The features of the coatings were observed, and the coatings thickness was measured on the cross sections of the coated specimens by a Tescan Mira 4 scanning electron microscope (Tescan, Brno, Czech Republic).

The surface hardness and Young's modulus of the UN and PN materials were measured using the instrumented indentation test by the Zwick ZHU 2.5 hardness tester (Zwick, Brno, Czech Republic). For the instrumented indentation test, the test force of 49.81 N and dwell time of 12 s were used. These measurements were performed ten times and the averages were used as the data. The microhardness profiles of the nitrided layers were obtained in the range from the surfaces to 0.5 mm depth on the polished cross sections. The microhardness was measured using the Vickers method on the Leco AMH55 hardness tester (Leco Corporation, St. Joseph, MI, USA). For the microhardness measurement the test force of 0.981 N and dwell time of 12 s were used. The microhardness measurement was carried out three times at each depth and their average values were used as the experimental values. The depth of nitrided layer was evaluated base on the results of measurement microhardness profile according to ISO-18203 [22]. The nanohardness and Young's modulus of the AlCrBN and AlCrVN coatings were measured 10 times using a nano-indenter, and their averages were used as the data.

The adhesion of the coatings was evaluated by the scratch test using the Universal Mechanical Tester 3 (UMT-3) and by the Rockwell indentation test using an Indentec HRC 8150 LK hardness tester (Zwick/Roell GmbH, Ulm, Germany). For these tests, the Rockwell diamond indenter with a rounded tip in radius of 200 µm was used. In the scratch test, the sliding speed was 0.17 mm/s, and the moving distance was 10 mm. The normal load was linearly increased from 5 to 150 N and 200 N for only-coated and duplex-coated materials, respectively. The critical loads were determined using the signal of acoustic emission and friction coefficient. For the Rockwell indentation test the test force was 1471 N and dwell time was 12 s. The tests were performed three times for all coatings. The friction-wear qualities of the materials were assessed using the linearly reciprocating ball-on-flat method. Friction-wear experiments were conducted at room temperature in air using a UMT-3 tester (Bruker, Karlsruhe, Germany) with no lubrication. The opposing substance was tungsten carbide balls with a diameter of 6.35 mm. The test load was set to 10 N. Table 3 depicts the test circumstances.

Table 3. Conditions of the wear tests.

Stroke Length (mm)	Oscillating Frequency (Hz)	Test Duration (s)	Ambient Temperature (°C)	Relative Humidity (%)	Lubrication
10	3.5	2000	22 ± 0.5	40–60	None applied

The features of wear tracks were evaluated using an Olympus DSX 500i optical microscope (Olympus, Tokyo, Japan) after the friction-wear testing. The wear track profiles were collected using an absolute inductive position sensor and a Talysurf CLI 1000 stylus profilometer (Taylor Hobson Ltd., Leicester, England). At eleven points along the wear track, profile measurements were taken. The mean cross-sectional profile was calculated from the measured profiles, and Talysurf Platinum software (v5, Taylor Hobson Ltd., Leicester, England) was used to calculate the area and depth of worn track cross sections. From the cross-sectional area and stroke length, the total volume of material lost during sliding (wear volume) was estimated [23].

3. Results

3.1. Surface Roughness

The surface roughness (Ra) of the studied materials is depicted in Figure 1.

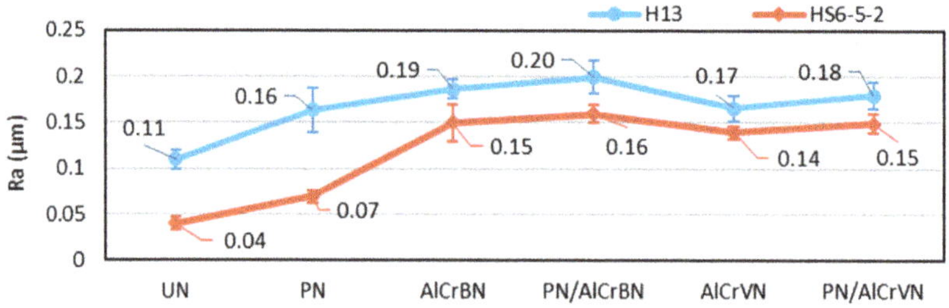

Figure 1. Surface roughness of materials.

In comparison with UN materials, the surface roughness Ra value of the PN material was 45% greater for H13 steel and 72% higher for HS6-5-2 steel. For steel H13, the surface roughness values were 50% to 70% higher than the UN material. Such deterioration is caused by a dedusting process, when the nitride cations bombard a material surface and subsequently atoms of various elements, being on a material surface are shot out [24]. The surface roughness of the coatings on steel HS6-5-2 was 245 percent to 295 percent higher than the UN material. Higher values of the surface roughness of the coating's parameters Ra and Rz have a detrimental impact on the coated part's friction, wear, and service life. Methods of surface treatment of coatings can be applied after coating deposition to minimize the values of the parameters Ra of surface roughness. Surface treatment of the coating with wet sandblasting and lap technologies reduces the value of the Ra roughness parameters of the coated surface [25].

3.2. Microstructures

The microstructure of H13 and HS6-5-2 steels is shown in Figure 2. Figures 3–5 demonstrate the microstructures found along the surfaces of PN, PN/AlCrBN, and PN/AlCrVN materials on cross sections.

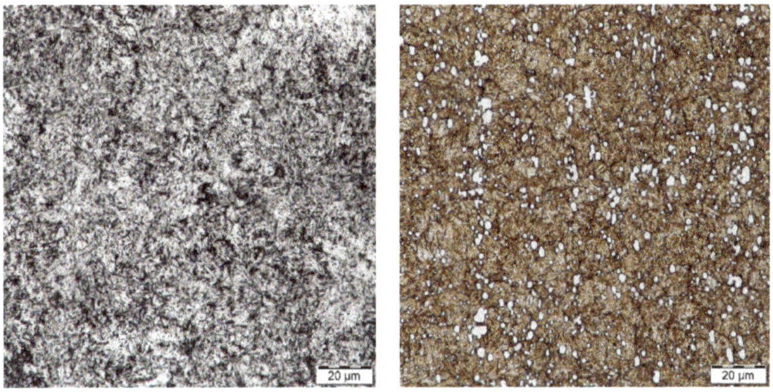

(**a**) (**b**)

Figure 2. Microstructure of steel (**a**) H13; (**b**) HS6-5-2 with magnification 1000×.

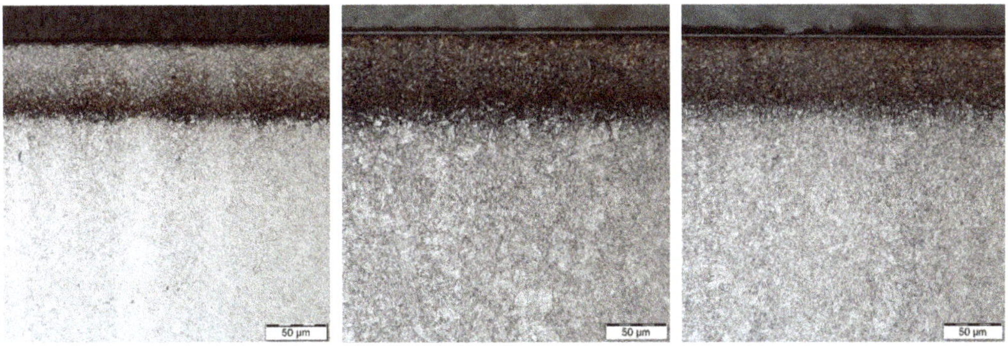

(**a**) (**b**) (**c**)

Figure 3. Microstructure optically observed with magnification 1000× on the steel H13 samples cross sections: (**a**) PN material; (**b**) PN/AlCrBN material; (**c**) PN/AlCrVN material.

(a) (b) (c)

Figure 4. Microstructure optically observed on the steel HS6-5-2 samples cross sections: (**a**) PN material; (**b**) PN/AlCrBN material; (**c**) PN/AlCrVN material.

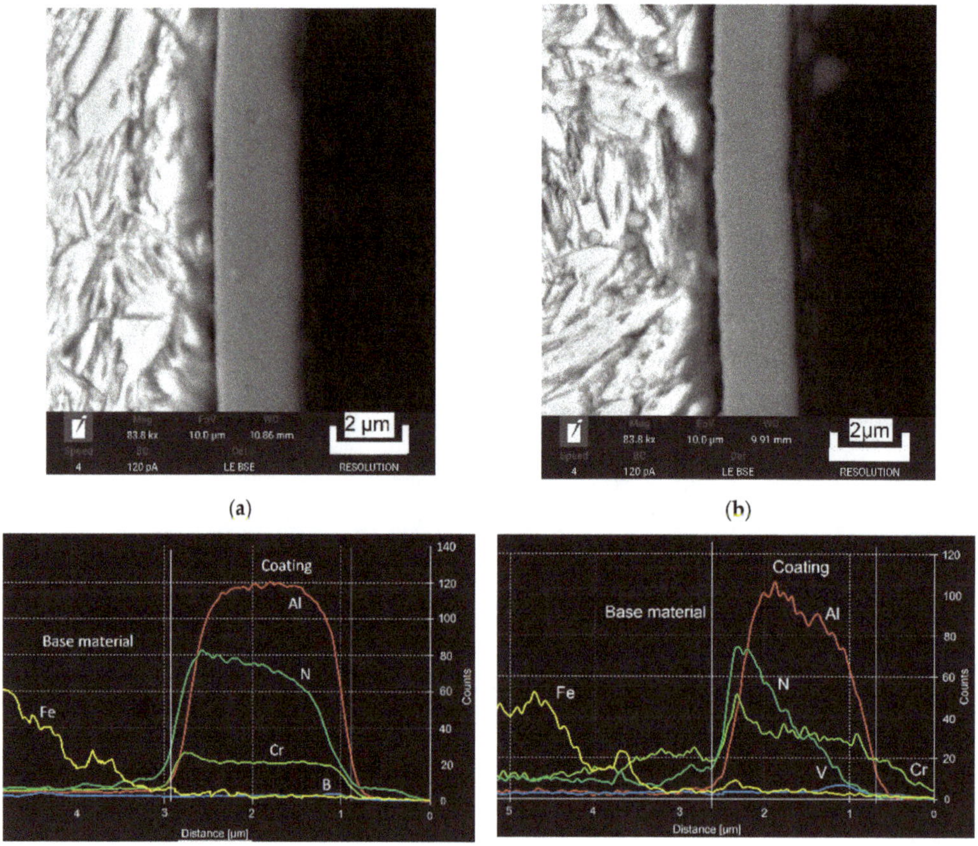

(a) (b)

(c) (d)

Figure 5. Features of the coatings on nitrided steel H13: (**a**) AlCrBN; (**b**) AlCrVN; (**c**) EDS line scan of AlCrBN coating; (**d**) EDS line of AlCrVN coating.

Steel H13 is a pearlitic steel with secondary carbides that is subledeburitic. The microstructure of H13 steel is tempered martensite following heat treatment, as shown in Figure 2a. The ledeburitic steel HS6-5-2 has a pearlitic structure. The martensite microstructure was detected in heated steel HS6-5-2, as shown in Figure 2b. The material HS6-5-2 has a fine-grained structure with carbides that are evenly distributed.

Figures 3a and 4a show the diffusion layers of the PN materials; nevertheless, no compound layer was generated. The duplex coatings, which include a thin coating and a thick nitrided layer, were formed on the cross sections of the duplex-coated materials (Figures 3b,c and 4b,c). The internal microstructures of PN materials, duplex-coated materials, and UN materials were identical, as illustrated in Figures 2–4. Due to their low conducting temperature, plasma nitriding and AlCrBN, AlCrVN coatings have little effect on the interior microstructures of steels.

Figure 5 depicts the characteristics of the AlCrBN and AlCrVN coatings as observed by SEM on cross sections.

On AlCrBN-coated materials, single coatings with a thickness of 2.0 μm were generated, as illustrated in Figure 5a. On the cross sections of AlCrVN-coated materials, a single coating of AlCrVN with a thickness of 1.8 μm was detected (Figure 5b).

3.3. Hardness

In Table 4, the surface hardness of UN and PN materials is compared.

Table 4. Surface hardness of the UN materials and PN materials.

Steel	Material	Hardness (HV 5)
H13	UN	502 ± 6
	PN	1120 ± 15
HS6-5-2	UN	874 ± 8
	PN	1335 ± 10

Plasma nitriding increased surface hardness by 123 percent for steel H13 and by 53 percent for steel HS6-5-2, as shown in Table 4. The microhardness profiles on the cross sections of PN materials are shown in Figure 6. The hardened layers were created by the plasma nitriding method, as shown in Figure 6. Microhardness profiles according to ISO-18203 [22] were used to determine the depth of the nitrided diffusion layers. The gradual decrease in hardness from the surface towards the core is due to the diffusion mechanism of the penetration of nitrogen atoms into the base material.

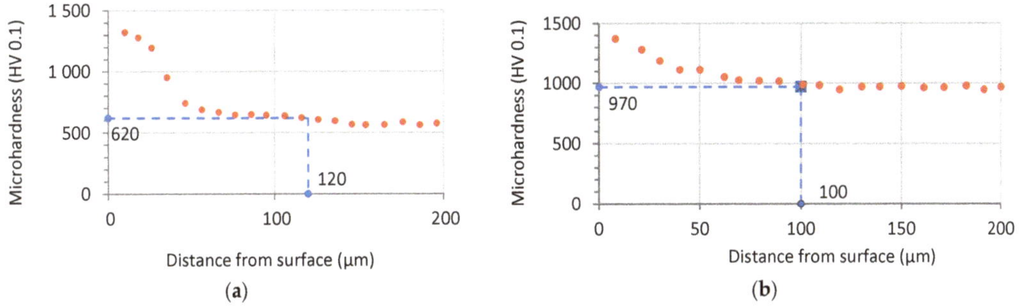

Figure 6. Microhardness profiles of the nitrided samples: (**a**) steel H13; (**b**) steel HS6-5-2.

Table 5 displays the maximum and minimum microhardness values, as well as the depth of nitrided layers.

Table 5. Properties of the nitrided layers.

Steel	Maximal Microhardness Value (HV 0.1)	Limit Microhardness Value (HV 0.1)	Case Depth (μm)
H13	1325 ± 65	620	120 ± 12
HS6-5-2	1375 ± 52	970	100 ± 10

Great variations in the mechanical characteristics of the two materials might lead to probable failure zones due to elastic and plastic incompatibility in only-coated systems, where a hard ceramic coating is put over a soft substrate material [16,21].

The nitrided layer was designed to improve the substrate load-bearing capacity in duplex-coated materials by providing a progressive transition in mechanical characteristics between the substrate and the hard ceramic coating [21,25]. The creation of the nitrided layer considerably increased the adhesive strength and tribological properties of ceramic coatings [17]. The ratio H/E^*, where $E^* = E/(1—v^2)$ and v is the Poisson's ratio, was commonly used to assess resistance to elastic strain to failure in the surface-contact mode, which is obviously significant for avoiding wear [26–28]. The resistance to plastic deformation in a surface contact system is measured using H^3/E^{*2} values [13,26,27,29].

Indentation hardness of UN and PN materials, nanohardness of AlCrBN and AlCrVN coatings evaluated by indentation test, and H/E^* and H^3/E^{*2} ratios of tested materials are shown in Table 6.

Table 6. Ratio H/E^* and H^3/E^{*2} of the tested materials.

Material-Steel	HIT (GPa)	E* (GPa)	$H/E^* \times 10^{-3}$	$H^3/E^{*2} \times 10^{-3}$ (GPa)
UN-H13	5.6 ± 0.4	230.8 ± 15	24 ± 3	3.3 ± 1.1
PN-H13	11.5 ± 0.6	224.2 ± 8	51 ± 5	30.3 ± 6.9
UN-HS6-5-2	9.5 ± 0.4	230.0 ± 10	41 ± 4	16.2 ± 3.5
PN-HS6-5-2	13.9 ± 0.5	237.4 ± 7	59 ± 4	47.7 ± 8
AlCrBN	40.9 ± 0.5	455.2 ± 9	90 ± 3	330.2 ± 25.2
AlCrVN	39.3 ± 0.5	438.3 ± 8	89 ± 3	315.5 ± 23.6

The highest H/E^* and H^3/E^{*2} ratios of 0.090 and 0.3302 GPa, respectively, were observed for the coating AlCrBN, as shown in Table 6. The values associated with the AlCrVN coating were slightly lower. The ceramic coatings had substantially higher values than the PN materials.

For steel HS6-5-2, the value of H/E^*, H^3/E^{*2} corresponding to the nitrided layer was about half and four times that of un-nitrided material. For steel H13, the nitride material enhanced the value of H/E^* and H^3/E^{*2} by nearly twice and ten times, respectively, as compared to un-nitrided material.

3.4. Adhesion Strength

One of the most essential properties of ceramic coatings is adhesion strength. It determined the service life of coating materials in a direct manner [30]. The Rockwell indentation and scratch test were frequently used to evaluate the adhesion strength of ceramic coatings to substrates.

The optical micrographs of Rockwell indentation of AlCrBN coatings and PN/AlCrBN duplex coatings on substrate steel H13 and steel HS6-5-2, respectively, are shown in Figures 7 and 8.

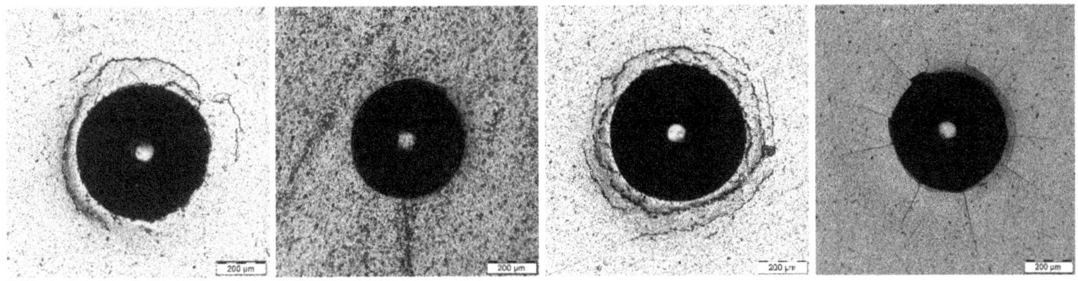

(**a**) (**b**) (**c**) (**d**)

Figure 7. The optical micrograph of Rockwell indentation of the coatings on substrate steel H13 with magnification 250×: (**a**) AlCrBN; (**b**) PN/AlCrBN; (**c**) AlCrVN; (**d**) PN/AlCrVN.

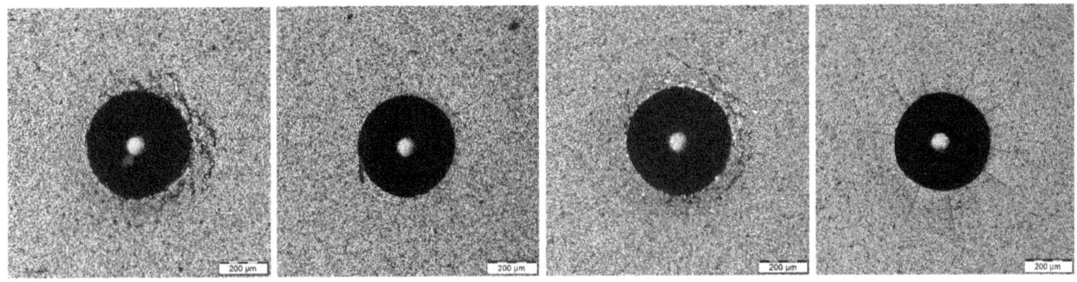

(**a**) (**b**) (**c**) (**d**)

Figure 8. The optical micrograph of Rockwell indentation of the coatings on substrate steel HS6-5-2 with magnification 250×: (**a**) AlCrBN; (**b**) PN/AlCrBN; (**c**) AlCrVN; (**d**) PN/AlCrVN.

The substrates un-nitrided materials of indentation were extruded to the edge and certain bulges were generated, as shown in Figure 7a,c and Figure 8a,c. Around the margin of the indentation, circular cracks through coatings and tiny spallation of coatings developed. The adhesion strength of AlCrBN only-coating and AlCrVN only-coating generated on substrate steel H13 and steel HS6-52 can be categorized as HF3 and HF2, respectively, according to the Rockwell indentation test VDI 3198 standard [31].

Because the coatings at the indentation edge rose during the loading process producing increased transverse shear stress and longitudinal tensile stress, and the AlCrBN and AlCrVN coatings exhibited high hardness and great brittleness, fractures formed under the compressive stress [32]. In comparison with substrates HS6-5-2, substrates H13 had larger indentation bulges due to their reduced hardness. The adhesion strength of the AlCrBN and AlCrVN coatings created on un-nitrided substrates steel H13 and steel HS6-52 was similar to HF2 when compared to the quality level of bonding strength according to Rockwell indentation test VDI 3198 standards. Figure 9 shows an EDS analysis of the neighborhood of indentation after the indentation test. An analysis of the presence of the two main elements that occur in the coating and the base material, Al and Fe, was performed. The occurrence of Fe was measured in the area of the peeled off part of the coating, which means that the peeling took place in the whole volume of the coating and not in its layers.

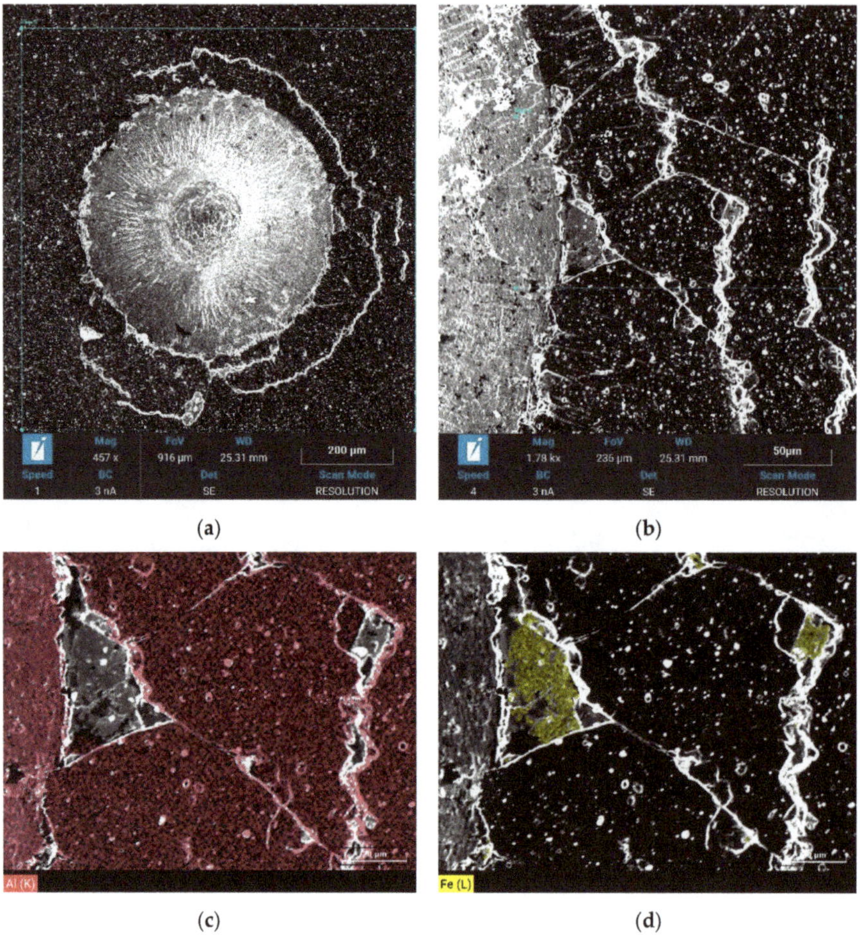

Figure 9. EDS analysis of cracks and the peeled AlCrVN coating: (**a**) surface feature of the indentation; (**b**) surface feature of the cracks and spallation; (**c**) distributions of Al; (**d**) distributions of Fe.

The diameter of the indentations (respectively plastic defamation) was larger in the duplex-coated materials than in the only-coated materials. It was discovered that the surface sink-in around the indentations had occurred. Figures 7b,d and 8b,d show a few cracks around the margin of the indentation on the surface of the duplex-coated materials. The adhesion strength of AlCrBN and AlCrVN coatings produced on nitrided steel H13 and steel HS6-52 substrates was comparable to that of HF1. As can be seen in Figures 7 and 8, the AlCrVN coatings had a greater number of cracks and longer cracks on their surface than the AlCrBN coatings. According to the findings, AlCrBN coatings have a higher toughness than AlCrVN coatings.

Figure 9a,b show SEM picture of crater after indentation test with EDS analyses of spallation. By analyzing the elements Fe and Al in Figure 9c,d, it was found that the coating peeled off in its entire thickness. Figure 10 depicts the critical loads of AlCrBN coating on un-nitrided substrate steel HS6-5-2, scratch optical micrographs, and the link between friction coefficient, acoustic emission, test load, and time test assessed by scratch.

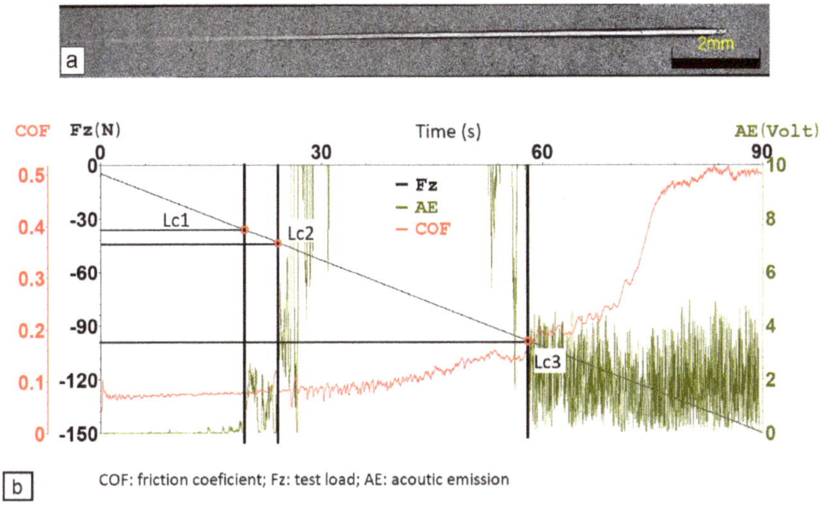

Figure 10. Critical loads of the coating AlCrBN on un-nitrided substrate of steel HS6-5-2 measured by scratch test: (**a**) optical micrographs of scratch scar; (**b**) relationship between the COF, AE, Fz and time test.

The critical load is defined as the least load at which a discernible failure occurs in a scratch test with progressive load [33]. The scratch test has three stages that correspond to three critical loads [30,32,34–36]. Each failure event, such as coating cracking and delamination, causes acoustic emission and a change in the coefficient of friction. The first AE peak and friction oscillation correspond to the first critical load Lc1, which is typically attributed to the first crack event. The delamination of coatings with substrate exposure determines the second critical load Lc2, which is usually connected with the adhesion strength between the coating and the substrate. The development of a quick increase in AE and random changes in the friction coefficient signal the start of delamination. The third critical load Lc3, which corresponds to the complete removal of a coating from the scratch groove, can be calculated using a microscopic study of the scratch tracks and a quick increase in coefficient friction. The second critical load, Lc2, is often seen as a symptom of coating adhesion failure [30].

The critical loads of AlCrBN and AlCrVN coatings produced on various substrates are shown in Figure 11.

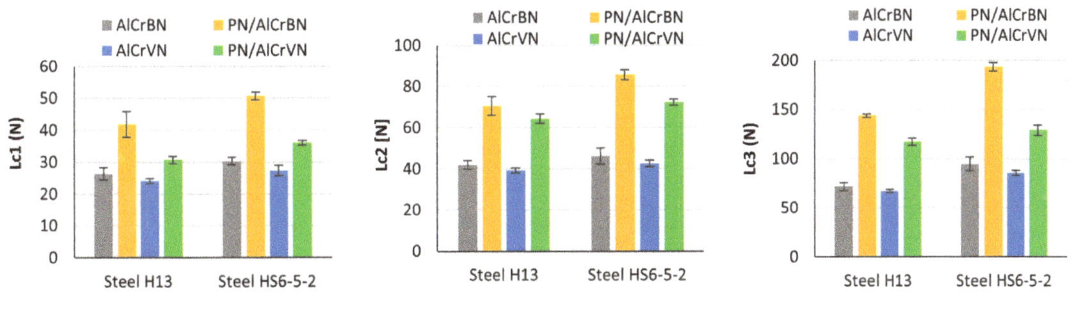

(**a**) (**b**) (**c**)

Figure 11. Critical loads of the AlCrBN and AlCrVN coatings formed on different substrates: (**a**) first critical load Lc1; (**b**) second critical load Lc2; (**c**) third critical load Lc3.

The AlCrBN nanocomposite coatings had a high hardness, but they had a lower compressive residual stress [14] and a lower internal stress of the coating system by >50% [15]. This improves the hardness [7,14,37] and adhesive strength of the AlCrBN coatings [7,27]. The film's resistance to cracking improves as the H^3/E^{*2} ratio rises [27]. As demonstrated in Table 6, the ratio H^3/E^{*2} of the AlCrBN coating was higher than that of the AlCrVN coating. As a result, the critical loads of the AlCrBN coating for substrates were higher than those of the AlCrVN coating, as shown in Figure 11.

Coatings formed on steel HS6-5-2 substrates had higher adherence than steel H13 substrates, and coatings formed on nitrided substrates had substantially higher adhesion than un-nitrided substrates. The findings indicated that the nitrided layer improved the substrate load-bearing capability by allowing for a progressive change in mechanical characteristics between the substrate and the hard coating, resulting in a significant increase in the adhesion strength of the AlCrBN and AlCrVN coatings.

3.5. Friction-Wear Properties

The relationship between the test period and the friction coefficients of all the investigated materials is depicted in Figure 12. Figure 12 shows that the UN steel H13 material has an unstable friction coefficient ranging from 0.42 to 0.66. To investigate the cause of the decrease in friction coefficient, three further wear tests were performed on UN material of steel H13. The first tribological test took 200 s to complete, corresponding to a friction coefficient of roughly 0.6. The second addition tribological test was completed in 400 s, equating to a friction coefficient of 0.42. The third tribological test took 1000 s to complete, equal to a friction coefficient of roughly 0.6. Figure 13 depicts the morphology of the wear tracks after additional tribological testing have been completed. The untreated sample has a morphology indicative of ploughing and oxidation, as seen in Figure 13. This is due to the fact that the H13 steel substrate has a lower hardness than the corresponding material (WC-Co 6 percent). The friction coefficient of two sliding contact surfaces is affected by the deposition of wear debris layers on wear track surfaces [38,39]. For the first addition wear test, the wear debris and sintered wear debris layers generated on the side of the wear track were detected (Figure 13a). Figure 13b shows how sintered wear debris layers cover practically the entire surface of the wear track, increasing the friction coefficient from roughly 0.6 to 0.42.

Figure 13c depicts the dispersed dispersion of thin wear debris layers. The worn surface was heavily oxidized. The removal of the wear debris layers from the wear track surface can be caused by a change in the surface of the counterpart material (ball). The ball was worn out during the test, and the surface of the balls that came into touch with the samples became flat. As a result of the ball ploughing, the wear debris layers are forced away from the wear track surface.

The tribological tests included two stages, referred to as the running-in stage and the stable stage, respectively, as shown in Figure 11. Except for un-nitrided steel H13, which started at a test time of 1000 s, the stable stages of tested materials began at around 400 s. For each test, the measured value of the friction coefficient is derived as the average of the stable region of the friction coefficient curve.

Table 7 shows the results of using three measured friction coefficients for tested materials to calculate the final values.

Table 7. Friction coefficient of tested materials.

Steel	UN	PN	AlCrBN	PN/AlCrBN	AlCrVN	PN/AlCrVN
H13	0.57 ± 0.05	0.63 ± 0.04	0.53 ± 0.02	0.50 ± 0.03	0.60 ± 0.03	0.62 ± 0.02
HS6-5-2	0.62 ± 0.04	0.64 ± 0.04	0.55 ± 0.02	0.56 ± 0.03	0.59 ± 0.03	0.57 ± 0.03

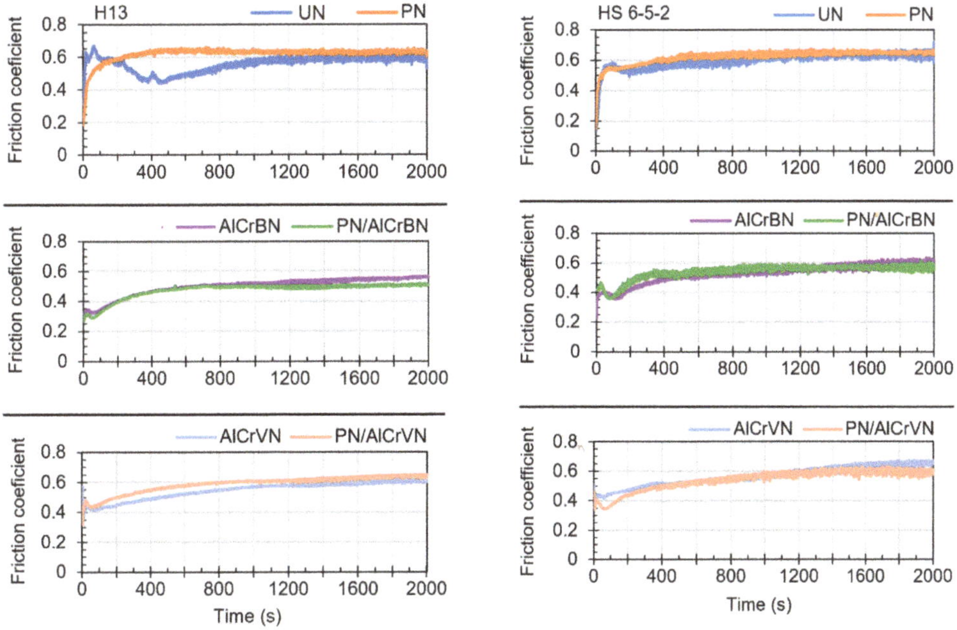

(a) (b)

Figure 12. Relationship between the test time and friction coefficients of the tested materials (**a**) Steel H13; (**b**) Steel HS6-5-2.

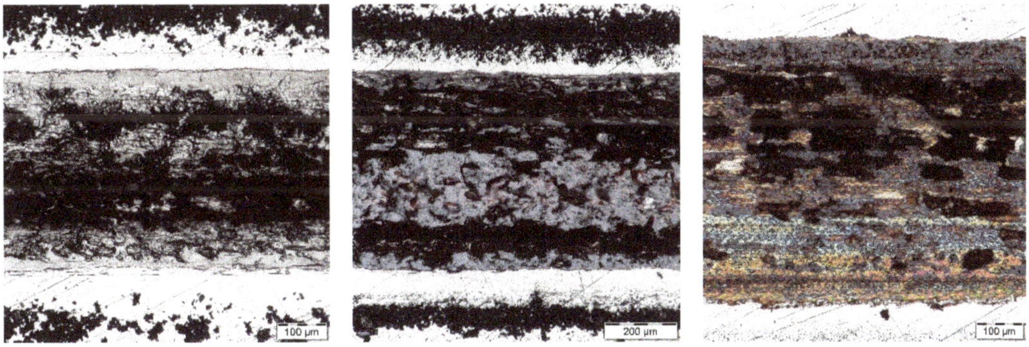

(a) (b) (c)

Figure 13. Morphology of the wear tracks on UN material of steel H13 after finishing addition tribological tests at time of: (**a**) 200 s; (**b**) 400 s; (**c**) 1000 s.

The characteristics and cross-sectional profiles of the wear tracks of the investigated samples steel H13 and HS6-5-2 are shown in Figures 14 and 15.

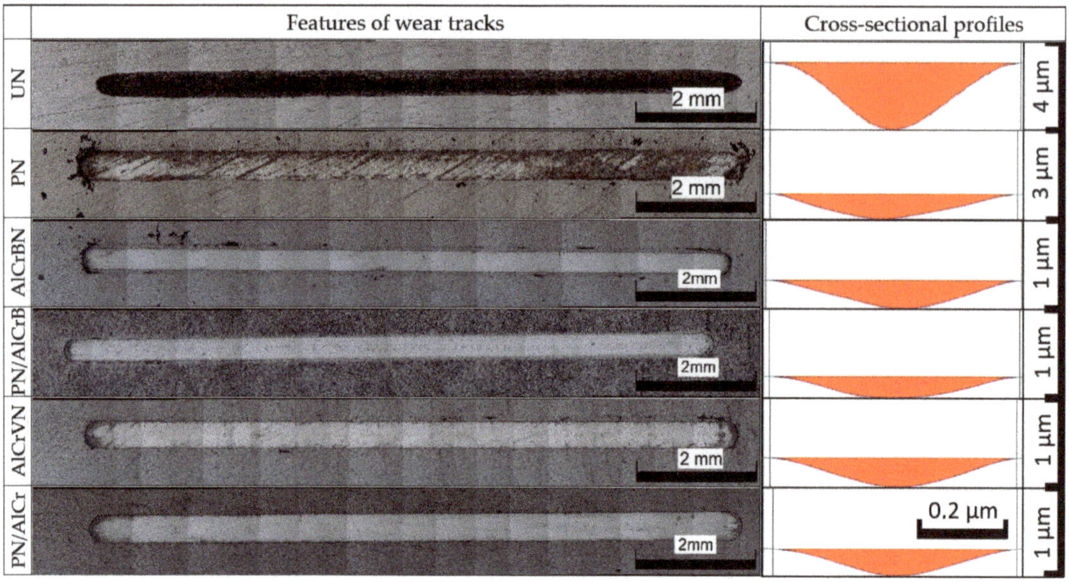

Figure 14. Features and cross-sectional of wear tracks of tested materials steel H13.

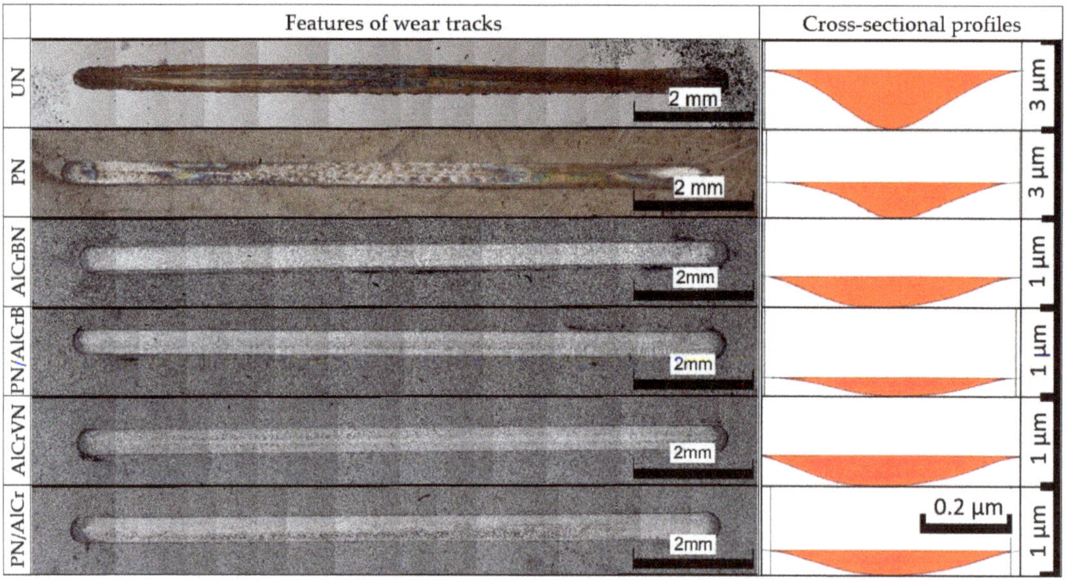

Figure 15. Features and cross-sectional of wear tracks of tested materials steel HS6-5-2.

The PN materials had the highest friction coefficient, as seen in Table 7 (0.63 and 0.64 for steel H13 and HS6-5-2, respectively). The friction coefficient values for the UN materials and AlCrVN coated materials varied between 0.6 and 0.7. The AlCrBN coated materials had the lowest friction coefficient, which was around 0.53.

Surface contact area, surface shear strength, surface hardness, and surface roughness all affect the friction coefficient [17,40]. The wear debris layers that can accumulate during a tribological test can also affect the friction coefficient [38,39]. The friction coefficient is influenced by the material's microstructure [41–43]. Even if the hardness is raised, the wear resistance is enhanced [17]. Friction coefficients do not always decrease.

Coating failures did not occur, and no substrate exposure was discovered on the wear tracks of the coated materials, as illustrated in Figures 14 and 15.

The wear volumes and depths of the tested materials of steel H13 and HS6-5-2 steel are shown in Figures 15 and 16, respectively.

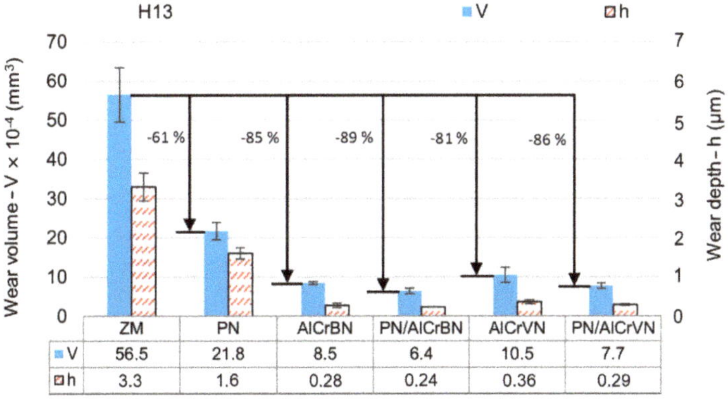

Figure 16. Wear volumes and wear depth of the tested materials of steel H13.

As shown in Figures 16 and 17, the wear depths of all the coated materials were lower than the coating thickness. The wear resistance of surface materials is generally related to the surface hardness, and then to the H/E* and H^3/E^{*2} ratios. The hardness and H/E* and H^3/E^{*2} ratios of the treated materials were higher than those of the untreated materials, as shown in Tables 4–6. As a result, for both steel H13 and HS6-5-2, the wear volumes of all treated materials (PN, only-coated, and duplex-coated) were lower than the steel substrate. PN materials exhibited 61 percent and 44 percent lower wear volumes for steel H13 and HS6-5-2, respectively, when compared to untreated materials. Wear volume reductions ranged from 81 percent to 89 percent for coated steel H13 materials and 66 percent to 80 percent for coated steel HS6-5-2 materials.

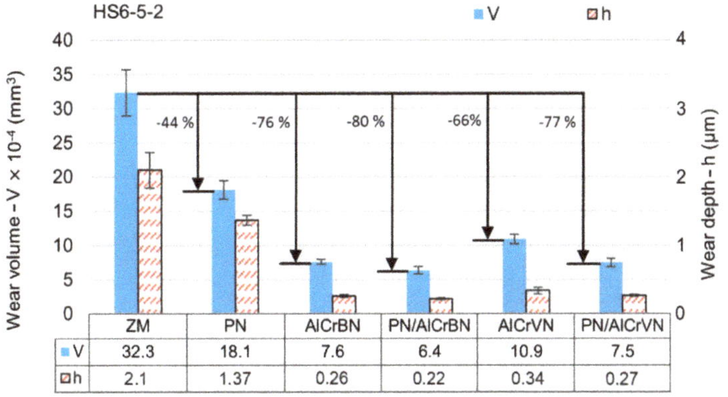

Figure 17. Wear volumes and wear depth of the tested materials of steel HS6-5-2.

The development of nitrided layers reduced plastic deformation caused by the coatings and raised the H/E* and H^3/E^{*2} ratios of the duplex-coated materials. In all situations, duplex-coated materials had lower wear volume values by 16% to 31% as compared to only-coated materials, according to the results of wear volume measurement. The AlCrBN coating displayed excellent hardness, low compressive residual stress, good toughness, and good wear resistance by combining the effects of solid solution hardening, grain size refinement (Hall–Petch hardening), and development of nanocomposite structure by boron addition [9,14,15]. The production of V_2O_5, which can operate as a liquid lubricant and reduce friction coefficients from 0.6–0.8 to 0.2–0.3 at high temperatures (700 °C), was one of the key benefits of the vanadium addition [11,44]. In fact, for steel H13 and HS6-5-2, the wear volume of the AlCrVN only-coated material was 24% and 43% higher than that of the AlCrBN only-coated material, respectively. For steel H13 and HS6-5-2, the PN/AlCrVN duplex-coated material had a 20 percent and 18 percent higher wear volume than the PN/AlCrBN duplex-coated material, respectively.

4. Conclusions

The purpose of this investigation was to see how surface roughness, mechanical characteristics, and friction-wear parameters of H13 and HS6-5-2 steel were affected by a hybrid treatment that included plasma nitriding and AlCrVN, AlCrBN ceramic coating. The effects of adding boron and vanadium to the AlCrN coating on mechanical, adhesion strength, and friction-wear properties were investigated.

The harder layers generated by plasma nitriding significantly increased the adherence of the AlCrBN and AlCrVN coatings. The AlCrBN coatings had greater adherence than the AlCrVN coatings.

H/E^* and H^3/E^{*2} ratios of 0.090 and 0.3302 GPa, respectively, were highest in the AlCrBN coating. The AlCrVN coating's H/E^* and H^3/E^{*2} values, which were 0.089 and 0.3155, respectively, were a little lower than the AlCrBN coating's. The coatings had substantially higher values than the PN materials. The value of H/E^*, H^3/E^{*2} corresponding to the nitrided layer in steel HS6-5-2 was roughly half and four times that of un-nitrided material, respectively.

Steels with a hybrid surface treatment consisting of nitriding and AlCrBN or AlCrVN coatings have considerably better friction-wear qualities than steels with simply coatings or nitrided layers.

The AlCrBN coating was harder, adhered better, had a lower friction coefficient, and was more resistant to wear than the AlCrVN coating. The results suggested that adding boron to the AlCrN system multicomponent coatings increased their mechanical and friction-wear qualities significantly.

Author Contributions: Conceptualization, H.C.N. and Z.J.; methodology, H.C.N., Z.J., E.S. and Z.S.; formal analysis, H.C.N.; investigation, H.C.N. and Z.J., J.S., J.M.; resources, H.C.N., J.S., J.M., D.D. and Q.D.T.; writing—original draft preparation, H.C.N., Z.J. and D.D., writing—review and editing, H.C.N., Z.J., Z.P., Q.D.T. and J.P.; supervision, E.S. and Z.P. All authors have read and agreed to the published version of the manuscript.

Funding: The work presented in this paper has been supported by the specific research project 2020 "SV20-216"at the Department of Mechanical Engineering, University of Defence in Brno and the Project for the Development of the Organization "DZRO Military autonomous and robotic systems" and the Slovak Research and Development Agency under contract No. APVV-15-0710.

Institutional Review Board Statement: Not applicable.

Informed Consent Statement: Not applicable.

Data Availability Statement: Data are contained within the article.

Acknowledgments: The authors acknowledge the infrastructure and support of Department of Industrial Engineering and Information Systems, Faculty of Management and Economics, Tomas Bata University in Zlín to carry out this research.

Conflicts of Interest: The authors declare no conflict of interest.

References

1. Hrbáčková, L.; Stojanović, A.; Tuček, D.; Hrušecká, D. Environmental Aspects of Product Life Cycle Management and Purchasing Logistics: Current Situation in Large and Medium-Sized Czech Manufacturing Companies. *Acta Polytech. Hung.* **2019**, *16*, 79–94. [CrossRef]
2. Mayrhofer, P.H.; Rachbauer, R.; Holec, D.; Rovere, F.; Schneider, J.M. Protective Transition Metal Nitride Coatings. In *Comprehensive Materials Processing*; Elsevier: Amsterdam, The Netherlands, 2014; Volume 4, pp. 355–388. [CrossRef]
3. Kalss, W.; Reiter, A.; Derflinger, V.; Gey, C.; Endrino, J.L. Modern Coatings in High Performance Cutting Applications. *Int. J. Refract. Met. Hard Mater.* **2006**, *24*, 399–404. [CrossRef]
4. Kawate, M.; Hashimoto, A.K.; Suzuki, T. Oxidation Resistance of Cr1-XAlxN and Ti1-XAlxN Films. *Surf. Coat. Technol.* **2003**, *165*, 163–167. [CrossRef]
5. Spain, E.; Avelar-Batista, J.C.; Letch, M.; Housden, J.; Lerga, B. Characterisation and Applications of Cr-Al-N Coatings. *Surf. Coat. Technol.* **2005**, *200*, 1507–1513. [CrossRef]
6. Franz, R.; Sartory, B.; Kaindl, R.; Tessadri, R.; Reiter, A.; Derflinger, V.H.; Polcik, P.; Mitterer, C. High-Temperature Tribological Studies of Arc- Evaporated Al x Cr 1-x N Coatings. In Proceedings of the 16th International Plansee Seminar, Reutte, Austria, 30 May–3 June 2005.
7. Sato, T.; Yamamoto, T.; Hasegawa, H.; Suzuki, T. Effects of Boron Contents on Microstructures and Microhardness in CrxAlyN Films Synthesized by Cathodic Arc Method. *Surf. Coat. Technol.* **2006**, *201*, 1348–1351. [CrossRef]
8. Musil, J. Hard and Superhard Nanocomposite Coatings. *Surf. Coat. Technol.* **2000**, *125*, 322–330. [CrossRef]
9. Tritremmel, C.; Daniel, R.; Rudigier, H.; Polcik, P.; Mitterer, C. Mechanical and Tribological Properties of Al-Ti-N/Al-Cr-B-N Multilayer Films Synthesized by Cathodic Arc Evaporation. *Surf. Coat. Technol.* **2014**, *246*, 57–63. [CrossRef]
10. Franz, R.; Neidhardt, J.; Sartory, B.; Kaindl, R.; Tessadri, R.; Polcik, P.; Derflinger, V.H.; Mitterer, C. High-Temperature Low-Friction Properties of Vanadium-Alloyed AlCrN Coatings. *Tribol. Lett.* **2006**, *23*, 101–107. [CrossRef]
11. Iram, S.; Cai, F.; Wang, J.; Zhang, J.; Liang, J.; Ahmad, F.; Zhang, S. Effect of Addition of Mo or V on the Structure and Cutting Performance of AlCrN-Based Coatings. *Coatings* **2020**, *10*, 298. [CrossRef]
12. Franz, R.; Neidhardt, J.; Kaindl, R.; Sartory, B.; Tessadri, R.; Lechthaler, M.; Polcik, P.; Mitterer, C. Influence of Phase Transition on the Tribological Performance of Arc-Evaporated AlCrVN Hard Coatings. *Surf. Coat. Technol.* **2009**, *203*, 1101–1105. [CrossRef]
13. Dinesh Kumar, D.; Kumar, N.; Kalaiselvam, S.; Dash, S.; Jayavel, R. Substrate Effect on Wear Resistant Transition Metal Nitride Hard Coatings: Microstructure and Tribo-Mechanical Properties. *Ceram. Int.* **2015**, *41*, 9849–9861. [CrossRef]
14. Tritremmel, C.; Daniel, R.; Lechthaler, M.; Rudigier, H.; Polcik, P.; Mitterer, C. Microstructure and Mechanical Properties of Nanocrystalline Al-Cr-B-N Thin Films. *Surf. Coat. Technol.* **2012**, *213*, 1–7. [CrossRef]
15. Beutner, M.; Lümkemann, A.; Jilek, M.; Bloesch, D.; Cselle, T.; Welzel, F.; Ag, P. High Speed Gear Hobbing with Customized AlCrBN Coatings Coating of Fly-Hobbing Teeth LACS ® Lateral ARC and Central Sputtering Doping AlCrN with Boron by LACS ® Hobbing with Boron Doped AlCr (Ti) N. In Proceedings of the 16th International Conference on Plasma Surface Engineering, Garmisch-Partenkirchen, Germany, 17–21 September 2018.
16. Matthews, A.; Leyland, A. Hybrid Techniques in Surface Engineering. *Surf. Coat. Technol.* **1995**, *71*, 88–92. [CrossRef]
17. Morita, T.; Inoue, K.; Ding, X.; Usui, Y.; Ikenaga, M. Effect of Hybrid Surface Treatment Composed of Nitriding and DLC Coating on Friction-Wear Properties and Fatigue Strength of Alloy Steel. *Mater. Sci. Eng. A* **2016**, *661*, 105–114. [CrossRef]
18. Joska, Z.; Kadlec, J.; Hrubý, V.; Mrázková, T.; Maňas, K. Characteristics of Duplex Coating on Austenitic Stainless Steel. *Key Eng. Mater.* **2011**, 255–258. [CrossRef]
19. Prochazka, J.; Pokorny, Z.; Dobrocky, D. Service Behavior of Nitride Layers of Steels for Military Applications. *Coatings* **2020**, *10*, 975. [CrossRef]
20. Slany, M.; Sedlak, J.; Zouhar, J.; Zemcik, O.; Chladil, J.; Jaros, A.; Kouril, K.; Varhanik, M.; Majerik, J.; Barenyi, I.; et al. Material and Dimensional Analysis of Bimetallic Pipe Bend with Defined Bending Radii. *Teh. Vjesn.* **2021**, *28*, 974–982. [CrossRef]
21. TSCHIPTSCHIN, A.P. *Duplex Coatings. Encyclopedia of Tribology*; Springer Science: Berlin/Heidelberg, Germany, 2013; pp. 794–799. [CrossRef]
22. ISO-18203. *Steel—Determination of the Thickness of Surface-Hardened Layers*; International Organization for Standardization: Geneva, Switzerland, 2016.
23. ASTM G133-02. *Standard Test Method for Linearly Reciprocating Ball-on-Flat Sliding Wear*; ASTM International: West Conshohocken, PA, USA, 2002.
24. Klanica, O.; Svoboda, E.; Joska, Z. Changes of the Surface Texture after Surface Treatment HS6-5-2-5 Steel. *Manuf. Technol. J.* **2015**, *15*, 47–53. [CrossRef]
25. Krbaťa, M.; Majerík, J.; Barényi, I.; Mikušová, I.; Kusmič, D. Mechanical and tribological features of the 90MnCrV8 steel after plasma nitriding. *Manuf. Technol. J.* **2019**, *19*, 238–242. [CrossRef]
26. Bell, T.; Dong, H.; Sun, Y. Realising the Potential of Duplex Surface Engineering. *Tribol. Int.* **1998**, *31*, 127–137. [CrossRef]
27. Leyland, A.; Matthews, A. On the Significance of the H/E Ratio in Wear Control: A Nanocomposite Coating Approach to Optimised Tribological Behaviour. *Wear* **2000**, *246*, 1–11. [CrossRef]
28. Musil, J.; Jirout, M. Toughness of Hard Nanostructured Ceramic Thin Films. *Surf. Coat. Technol.* **2007**, *201*, 5148–5152. [CrossRef]

29. Wang, Z.W.; Li, Y.; Zhang, Z.H.; Zhang, S.Z.; Ren, P.; Qiu, J.X.; Wang, W.W.; Bi, Y.J.; He, Y.Y. Friction and Wear Behavior of Duplex-Treated AISI 316L Steels by Rapid Plasma Nitriding and (CrWAlTiSi)N Ceramic Coating. *Results Phys.* **2021**, *24*, 104132. [CrossRef]
30. Brizmer, V.; Kligerman, Y.; Etsion, I. The Effect of Contact Conditions and Material Properties on the Elasticity Terminus of a Spherical Contact. *Int. J. Solids Struct.* **2006**, *43*, 5736–5749. [CrossRef]
31. Zhang, S.; Wang, L.; Wang, Q.; Li, M. A Superhard CrAlSiN Superlattice Coating Deposited by Multi-Arc Ion Plating: I. Microstructure and Mechanical Properties. *Surf. Coat. Technol.* **2013**, *214*, 160–167. [CrossRef]
32. Vidakis, N.; Antoniadis, A.; Bilalis, N. The VDI 3198 Indentation Test Evaluation of a Reliable Qualitative Control for Layered Compounds. *J. Mater. Process. Technol.* **2003**, *143–144*, 481–485. [CrossRef]
33. Tian-Shun, D.; Ran, W.; Guo-Lu, L.; Ming, L. Failure Mechanism and Acoustic Emission Signal Characteristics of Coatings under the Condition of Impact Indentation. *High Temp. Mater. Process.* **2019**, *38*, 601–611. [CrossRef]
34. ASTM C1624-05. *Standard Test Method for Adhesion Strength and Mechanical Failure Modes Of Ceramic Coatings By Quantitative Single Point Scratch Testing*; ASTM International, 100 Barr Harbor Drive, PO Box C700: West Conshohocken, PA, USA, 2005. [CrossRef]
35. Shugurov, A.; Akulinkin, A.; Voronov, A. Investigation of Adhesive Behavior of Ti-Al-N/Ti-Al Multilayers by Scratch Testing. *AIP Conf. Proc.* **2018**, *2051*, 020282. [CrossRef]
36. Wang, L.; Zhang, S.; Chen, Z.; Li, J.; Li, M. Influence of Deposition Parameters on Hard Cr-Al-N Coatings Deposited by Multi-Arc Ion Plating. *Appl. Surf. Sci.* **2012**, *258*, 3629–3636. [CrossRef]
37. Chen, W.; Zheng, J.; Lin, Y.; Kwon, S.; Zhang, S. Comparison of AlCrN and AlCrTiSiN Coatings Deposited on the Surface of Plasma Nitrocarburized High Carbon Steels. *Appl. Surf. Sci.* **2015**, *332*, 525–532. [CrossRef]
38. Shtansky, D.V.; Sheveiko, A.N.; Petrzhik, M.I.; Kiryukhantsev-Korneev, F.V.; Levashov, E.A.; Leyland, A.; Yerokhin, A.L.; Matthews, A. Hard Tribological Ti-B-N, Ti-Cr-B-N, Ti-Si-B-N and Ti-Al-Si-B-N Coatings. *Surf. Coat. Technol.* **2005**, *200*, 208–212. [CrossRef]
39. Zmitrowicz, A. Wear Debris: A Review of Properties and Constitutive Models. *J. Theor. Appl. Mech.* **2005**, *43*, 3–35.
40. Shi, H.; Du, S.; Sun, C.; Song, C.; Yang, Z.; Zhang, Y. Behavior of Wear Debris and Its Action Mechanism on the Tribological Properties of Medium-Carbon Steel with Magnetic Field. *Materials* **2018**, *12*, 45. [CrossRef]
41. Hutchings, I.M. *Tribology: Friction and Wear of Engineering Materials*; Edward Arnold: London, UK, 1992; Volume 13. [CrossRef]
42. Yang, Q.; Senda, T.; Ohmori, A. Effect of Carbide Grain Size on Microstructure and Sliding Wear Behavior of HVOF-Sprayed WC-12% Co Coatings. *Wear* **2003**, *254*, 23–34. [CrossRef]
43. Usmani, S.; Sampath, S.; Houck, D.L.; Lee, D. Effect of Carbide Grain Size on the Sliding and Abrasive Wear Behaviour of Thermally Sprayed WC-Co Coatings. *Tribol. Trans.* **1997**, *40*, 470–478. [CrossRef]
44. Franz, R. High-Temparature, Low-Friction Properties of the Vanadium Alloyed AlCrN Coatings. *Coaitngs* **2007**, *10*, 298. [CrossRef]

materials

MDPI

Article

The Influence of Heat Treatment on Low Cycle Fatigue Properties of Selectively Laser Melted 316L Steel

Janusz Kluczyński [1,*], Lucjan Śnieżek [1], Krzysztof Grzelak [1], Janusz Torzewski [1], Ireneusz Szachogłuchowicz [1], Artur Oziębło [2], Krzysztof Perkowski [2], Marcin Wachowski [1] and Marcin Małek [3]

1 Faculty of Mechanical Engineering, Institute of Robots & Machine Design, Military University of Technology, 2 Gen. S. Kaliskiego St., 00-908 Warsaw, Poland; lucjan.sniezek@wat.edu.pl (L.Ś.); krzysztof.grzelak@wat.edu.pl (K.G.); janusz.torzewski@wat.edu.pl (J.T.); ireneusz.szachogluchowicz@wat.edu.pl (I.S.); marcin.wachowski@wat.edu.pl (M.W.)

2 Department of Ceramics and Composites, Institute of Ceramics and Building Materials, 9 Postepu St., 02-676 Warsaw, Poland; artur.ozieblo@icimb.pl (A.O.); k.perkowski@icimb.pl (K.P.)

3 Faculty of Civil Engineering and Geodesy, Military University of Technology, 2 Gen. S. Kaliskiego St., 00-908 Warsaw, Poland; marcin.malek@wat.edu.pl

* Correspondence: janusz.kluczynski@wat.edu.pl

Received: 25 November 2020; Accepted: 15 December 2020; Published: 16 December 2020

check for updates

Abstract: The paper is a project continuation of the examination of the additive-manufactured 316L steel obtained using different process parameters and subjected to different types of heat treatment. This work contains a significant part of the research results connected with material analysis after low-cycle fatigue testing, including fatigue calculations for plastic metals based on the Morrow equation and fractures analysis. The main aim of this research was to point out the main differences in material fracture directly after the process and analyze how heat treatment affects material behavior during low-cycle fatigue testing. The mentioned tests were run under conditions of constant total strain amplitudes equal to 0.30%, 0.35%, 0.40%, 0.45%, and 0.50%. The conducted research showed different material behaviors after heat treatment (more similar to conventionally made material) and a negative influence of precipitation heat treatment of more porous additive manufactured materials during low-cycle fatigue testing.

Keywords: additive manufacturing; selective laser melting; mechanical properties; fatigue properties; heat treatment; hot isostatic pressing; 316L austenitic steel

1. Introduction

Additively manufactured (AM) parts are characterized by very distinctive properties, despite the fact that many different AM technologies where different types of materials are used exist. One of the most important features of parts obtained using AM is a layered structure of the manufactured parts, which significantly affects the anisotropy of mechanical properties during comparison of different building directions [1,2]. This phenomenon is also present in parts processed using laser-powder bed fusion (L-PBF) technologies. Good mechanical properties, a possibility of obtaining geometrically complex parts, and a large spectrum of available alloys allow for the dynamic growth of L-PBF technologies. Use of the AM is seen in aircraft [3,4], automotive [5], armament [6,7], and other solutions that require lightweight structures [8–11]. The development of new additively manufactured parts that are characterized by their special application needs to be supplied by a proper amount of knowledge about AM material properties and their behavior under different loading conditions. This kind of research project is in the scope of research of many scientific facilities [12–16].

One of the most common materials available for additive manufacturing is 316L steel, which in conventional-manufactured form, is dedicated for applications vulnerable to the adverse effects of chemical and biological factors because of its very good anti-corrosive properties. From a technological point of view, 316L steel belongs to the hard-to-cut materials group—mostly because of its austenitic structure. Additionally, the usage of this steel in medical applications often requires very complex geometry for exact parts. These two factors: applications in a corrosive environment and geometrical complexity, significantly affects material properties that are changing during operation for some specified time.

Despite that there are many available research results connected with mechanical properties of AM parts made of 316L steel, there is still a significant gap in the fatigue properties analysis.

The most significant group of available fatigue test results is connected with high-cycle testing (mostly Wohler's charts) [17–20], some works are connected with crack growth analysis of AM parts [21–23], but the smallest number of available research results concerns low-cycle fatigue properties, especially with some postprocessing connected with heat treatment or hot isostatic pressing (HIP).

Analysis of the Blinn et al. [24] research work revealed high anisotropy of AM material during fatigue testing, where test parts were manufactured in three different orientations. In samples manufactured vertically to the building plate, it was registered a higher defect tolerance on material damages, which led to higher endurance at lower stress amplitudes.

Mustafa et al. [25] analyzed an AM aluminum alloy, also considering low-fatigue cycle properties, in which the domination of extensive plastic damage beyond grain boundaries was discovered. Similar observations were registered by Romano et al. [26], where authors stated that defect size is the principal cause of variability in the fatigue resistance of the material, even in low cycle fatigue (LCF). Additionally, the authors obtained results where plasticity played an important role in the determination of the fatigue resistance of AM parts.

A different approach was suggested by Bressan et al. [27], in which the authors analyzed an influence of stress-relieving heat treatment on LCF properties of titanium alloys. During their tests, it was observed early sudden material's weakening, which was correlated with the formation of cracks in internal voids.

Based on two previously-mentioned research works [26,27], a significant influence of porosity in the material volume was stated. It is necessary to understand the reasons for porosity generation in AM parts [28,29]. It is possible to point to two characteristic parts of the AM material structure—porosity in the core of the material and near the outline borders. Regarding voids' presence in the material volume, an influence of layered structure and connection between fused layers of the material on mechanical properties was also analyzed. Shifeng et al. [30] determined how molten pool boundaries present in the material structure after AM processing affect the mechanical properties of manufactured parts. In the mentioned research, the authors indicated that molten pool boundaries have a significant impact on:

- microstructural slipping during loading;
- macroscopic plastic behavior;
- properties anisotropy;
- low ductility of SLM-processed parts,

On the other hand, Elangeswaran et al. [31] analyzed an influence of surface roughness on the fatigue properties, which indicated a high, negative effect of as-built samples (without surface machining). The main issue was related to unfused powder particles on the samples' surface, which were a kind of stress raisers.

Microfractographic analysis of the fracture surfaces of the samples allows for the description of the cracking mechanisms. Qualitative fractography plays a special role in explaining the cracking phenomena. It is possible to determine the origin of the crack, the nature of the cracking process. It is often possible to identify defects in material or manufacturing processes.

In this respect, fractal fractography works perfectly well, which deals with the complex aspects of fractures in materials.

Fractal microfractography allows for accurately determining the multidimensional course of the cracking process and is an indispensable element of product quality control [32,33].

During the literature review, it was difficult to find information about the LCF of 316L stainless steel processed using selective laser melting technology with additional analysis, including heat treatment and HIP. Moreover, an LCF analysis mentioned in cited literature included a low portion of microfracture analysis, which is very helpful to understand the damage mechanism of AM metallic parts.

Based on the authors' own previous research [14,15,34], it was a natural continuation of material analysis connected with the LCF of heat-treated and HIPped parts obtained using the selective laser melting technology (SLM). The main aim of the included research results was to determine the possibility of a void reduction in material volume and describe how it affects LCF properties.

2. Materials and Methods

2.1. Material

316L steel powder was used for the specimens manufacturing. The material was supplied by Carpenter Additive Company (Carpenter Additive, Widness, UK). It was gas-atomized in the argon atmosphere. Powder particles were characterized by a spherical shape in a diameter of 15–63 μm. Material's chemical composition is shown in Table 1.

Table 1. 316L steel chemical composition.

C	Cu	Mn	Si	O	P	S	N	Cr	Mo	Ni
					Weight (%)					
0.027	0.02	0.98	0.72	0.02	0.011	0.004	0.09	17.8	2.31	12.8

2.2. Additive Manufacturing Process Description

The shape of the samples (Figure 1) for fatigue testing was determined based on the recommendations of the ASTM E466 96 standard with considering the dimensions of the building volume of the AM system. For the research, the SLM 125HL system was used (SLM Solutions AG, Lubeck, Germany).

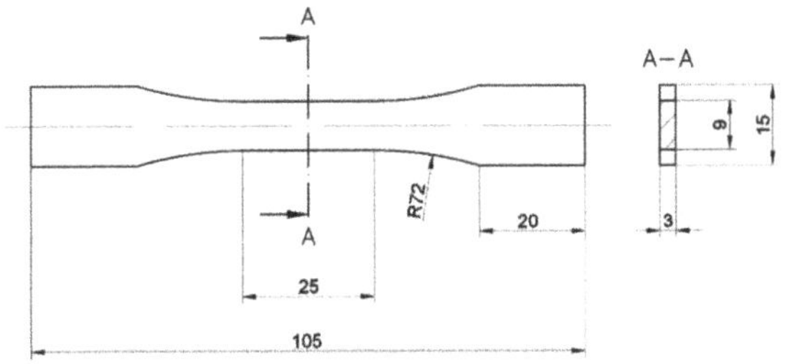

Figure 1. ASTM E466 96 samples shape.

Based on our previous research of the AM process [14,15,34], three-parameter groups were selected, as shown in Table 2. To be consistent with previously published papers—each sample group was named the same as before (Samples: S_01, S_17, S_30).

The S_01 samples were manufactured using producer's default settings, S_17 samples were manufactured with parameters which gave an increased porosity in [14], S_30 samples were made using almost three times the higher energy density than in default settings and based on

Di Wang et al.'s research [35] in which the authors reach good material properties with using mentioned process parameters.

Table 2. Parameter groups used for sample manufacturing.

Parameters Set	Laser Power L_P (W)	Exposure Velocity e_v (mm/s)	Hatching Distanc eh_d (mm)	Layer Thickness (mm)	Energy Density ρ_E (J/mm^3)
S_01	190	900	0.12	0.03	58.64
S_17	180	990	0.13	0.03	46.62
S_30	120	300	0.08	0.03	166.67

After SLM-processing, the samples were cut out from the build plate, and a sidewall of each sample was milled. The sharp edges of the sidewalls after milling were subjected to a roll-burnishing process to minimize the possibility of stresses arising in that area.

2.3. Heat Treatment

Each sample group was subjected to different heat treatment processes to reach different results, which allowed obtaining a higher amount of results and knowledge about the material. Hot isostatic pressing (HIP) was performed on a high-temperature press with furnace Nabertherm VHT 822—GR (Nabertherm, Lilienthal, Germany). Precipitation heat treatment (PHT) was done using the Nabertherm P300 furnace (Nabertherm, Lilienthal, Germany). Each heat treatment process is shown in Figure 2.

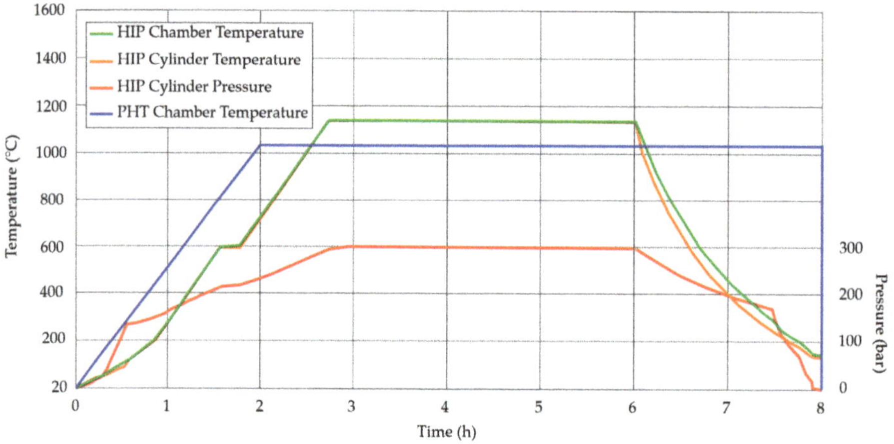

Figure 2. The course of heat treatment processes.

The HIP process was performed to reduce the volume porosity and remove the layered structure of the material after the SLM process. Furnace cooling after the HIP process could cause sigma phase generation in austenitic steel [36]—to avoid this and extend the research range, half of the HIPped samples were also subjected to PHT with water cooling, which allowed a significant reduction of the sigma phase formation phenomenon (it generates mostly between 700 and 850 °C).

2.4. Low Cycle Fatigue Testing

The scope of fatigue tests of SLM-processed 316L steel included testing of three series of samples produced with the use of three parameters selected during previous tests: S_01, S_17, and S_30. Samples subjected to HIP treatment were named as (S_01H and S_17H), samples subjected to precipitation heat treatment were named as (S_01P, S_17P, and S_30P), samples subjected to both types of heat treatment was named as (S_01HP and S_17HP). Samples geometry was made based on the recommendations of the standards [37–39]. The fatigue properties of the tested samples were determined based on the characteristic parameters of the hysteresis loops obtained at:

$$\frac{N}{N_f} = 0.5 \quad (1)$$

where:

N—current number of cycles;
N_f—number of cycles to failure.

During testing, the material fracture was assumed as the criterion of sample destruction. The value of the total strain amplitude ε_{ac} was changed sinusoidally with the strain ratio R = 0.1. Fatigue tests were performed with the frequency of load changes f = 0.8 Hz. The values of the total strain were determined based on the previously prepared static tension diagram, shown in [15]. The tests were carried out at five various levels of total strain amplitude: ε_{ac} = 0.30%; 0.35%, 0.40%; 0.45% and 0.50%. For each level, three samples were examined.

Fatigue tests were run on the Instron 8802 servohydraulic testing system and an Instron 2620–603 dynamic extensometer with a gauge length of 25 mm (Instron, Norwood, MA, USA).

2.5. Microscopical Analysis

For the fractography analysis, the fracture surfaces from failed specimens were first cut and mounted on an observation stage. The fracture surfaces low-cycle damaged samples were analyzed using the scanning electron microscope (SEM) Jeol JSM-6610 (Jeol, Tokyo, Japan).

3. Results and Discussion

3.1. Low Cycle Fatigue Properties Analysis

There is significant importance in determining the behavior of material exposed to strain-controlled low-cycle fatigue tests. Mechanisms present in this kind of loading cause local defect formation and consequently lead to the initiation of fatigue cracking. Fatigue strength testing of the SLM-processed materials is particularly justified due to the high possibility of structural imperfections occurrence, most often difficult to avoid without the use of additional postprocessing. A common example of damage in additive manufactured parts is the occurrence of cracks in the zone of increased porosity in the material structure. Such defects locally lower the fatigue strength, especially in thin-walled elements, which are more prone to the formation of porosity due to faster heat dissipation from the melting zone.

Some sample fractures were characterized by non-normative behavior caused by the samples buckling registered during hysteresis loops analysis. The scale of this phenomenon was significantly greater in the case of heat-treated parts. This negative effect was caused by the reduced yield strength of heat-treated samples. Consequently, such cases were not taken into account when compiling the test results.

The variations of stress amplitude σ_a with the load loops number are shown in Figure 3, respectively, for each sample group. Additive manufactured samples (S_01, S_17 and S_30) were compared with AM heat-treated parts (S_01H, S_01P, S_01HP, S_17H, S_17P, S_17HP and S_30). All results were compared to the conventionally made material (made of cold-rolled metal sheets—described as P_0 samples).

Concerning the additive manufactured and heat-treated samples, the total strain amplitudes σ_{ac}, are characterized by significantly smaller changes in the stress amplitude over the entire range of the number of load cycles N than in the case of non-heat-treated samples. This suggests that the AM material is less prone to cyclic weakening after heat treatment. The mentioned phenomenon is most visible in the case of samples S_01H, S_01P, and S_01HP. In these samples, the stage of cyclic weakening is transient. Additionally, in the range of 7–25% of N_f value (ε_{ac} = 0.30, 0.35, 0.40 and 0.45%) and 85–90% (ε_{ac} = 0.50%). The further course of the curves proves the material's cyclical stabilization. The samples S_01HP and S_17H showed the highest fatigue life.

Research results registered at the lowest value of the total strain amplitude ε_{ac} = 0.30%, shown total fatigue life at a level of 60% in comparison with conventionally manufactured 316L steel. The lowest fatigue life of the tested samples was achieved in samples S_17P (approx. 1550 cycles). It constitutes 16% of the P_0 sample's fatigue life.

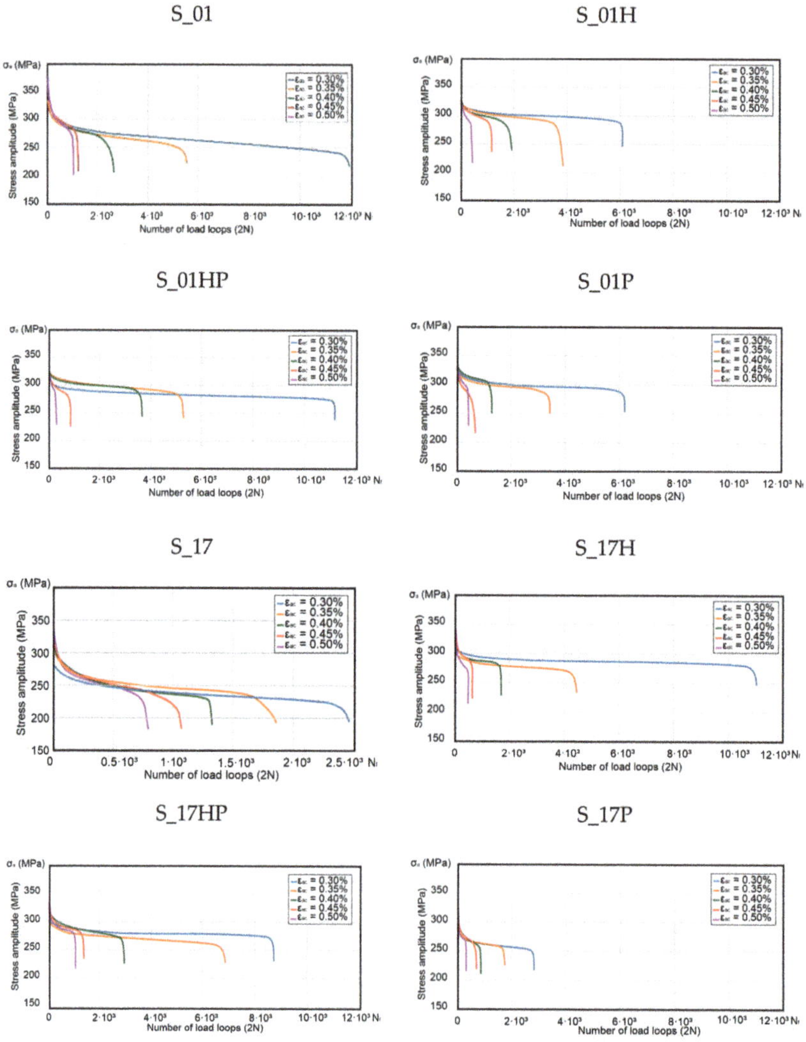

Figure 3. *Cont.*

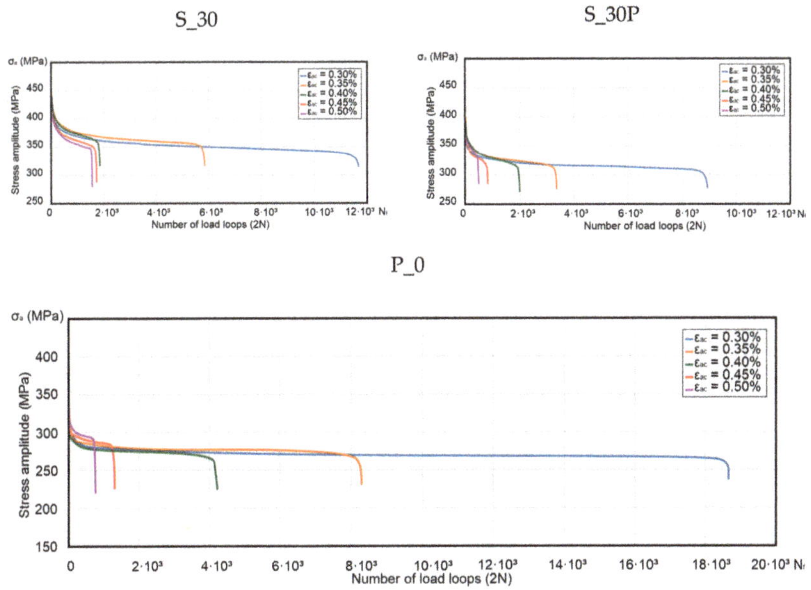

Figure 3. Variation of stress amplitude σ_a with the number of cycles for additively manufactured (AM), AM heat-treated parts and conventionally made material.

The similar fatigue life of heat-treated samples is noticeable in the samples tested directly after the additive manufacturing process. A significant difference was in the share of the cyclical weakening of the material, which in the case of heat-treated parts is significantly lower than in non-heat-treated samples. Analyzing Figure 3, it can be seen that achieving the same level of total strain amplitude for the conventionally made material (P_0) required a lower stress level than additively produced samples. This effect was due to increased fatigue-durability of conventionally manufactured samples, and at the same time, those samples did not exhibit greater strength at the same level of strain amplitudes as compared to AM samples.

As-built samples were characterized by increased stress levels in mid-life hysteresis loops. This phenomenon was strictly connected with the increased yield strength of as-built samples. In the case of all series of heat-treated samples, very similar courses of the mid-life hysteresis loops were observed during the tests carried out at the total strain amplitude of 0.30%. The coordinates of their peaks indicate the highest stress amplitude values occurrence in the S_30P samples. In the case of the remaining heat-treated samples, it was difficult to state a generalized regularity. The course and position of the peaks of the hysteresis loops (shown in Figure 4) obtained were very similar to the loop of the P_0 reference sample made of conventionally manufactured 316L steel.

To better understand differences between additive manufactured and conventionally made material, an example of the hysteresis loop for the P_0 samples in ε_{ac} = 0.3% strain condition for the peak/valley stressed of characteristic fatigue life cycles was shown in Figure 5. The course of the lower stress values did not change much, while the upper peaks with the number of cycles significantly decrease. The no. 1 cycle in the initial phase showed some irregularity, which may have been related to the compression of the sample in the testing machine holder. This cycle should not be analyzed. The shape of the hysteresis loops was fully regular during the cycles.

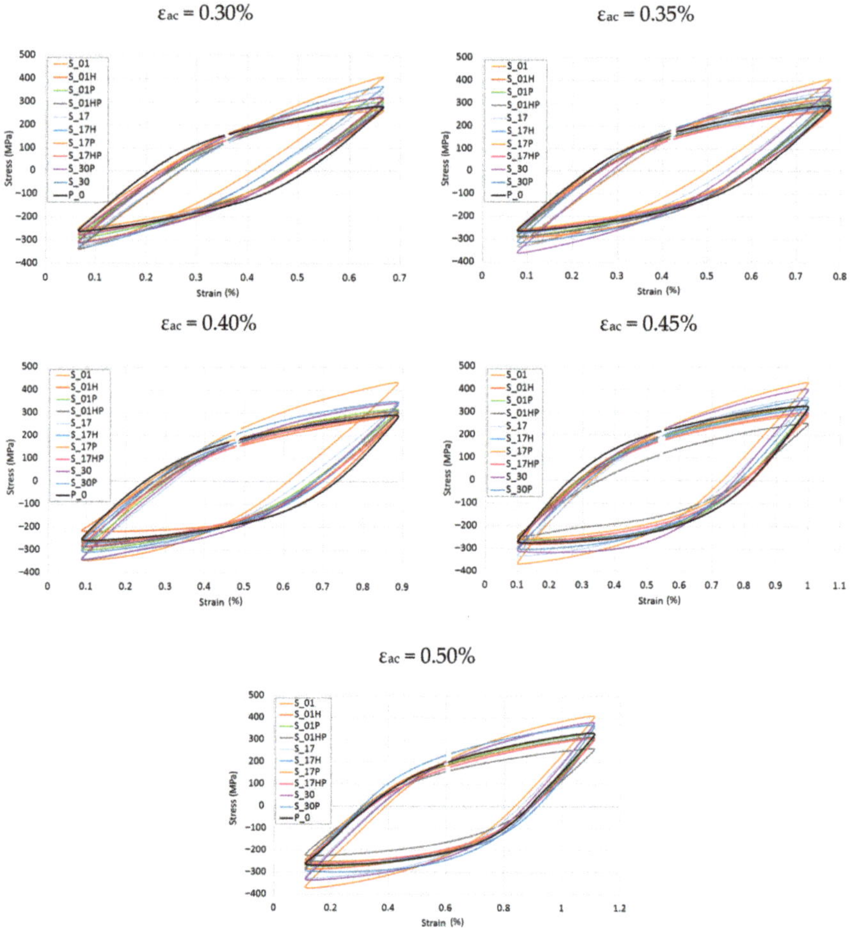

Figure 4. Mid-life hysteresis loops registered for total strain amplitudes ε_{ac} = 0.30, 0.35, 0.40, 0.45, and 0.50%.

Based on Figure 5 observations, it could be stated that the cyclic stress response of investigated specimens can give a qualitative description of cyclic softening phenomena. Similar observations under symmetrical strain control (R = −1) were made by H.S. Ho et al. [40], as well as softening and a fast stabilization presented by S. Romano et al. [26].

The main differences between materials after SLM processing and conventional made are also visible in Figure 3, where additive manufactured material, without any heat treatment weakened in the whole in the entire cycle range. At the same time—conventionally made material (and also additively manufactured after heat treatment) had a visible stabilization at some cycles range. This phenomenon could be connected with the material condition after additive manufacturing—similar to structure and properties after welding (visible melting pools in the structure, increased tensile strength, and decreased elongation at break).

The hysteresis loop surface area is a parameter providing information about material fatigue life as well as the amount of energy necessary for its destruction. Microlevel strains are irreversible plastic deformations, which are related to energy dissipation. It is the main factor causing material damage and the formation of fatigue microcracks. To facilitate measured values of dissipated energy by each sample, hysteresis loop areas were calculated using AutoCAD software (version 2020). The calculated values are shown in the chart (Figure 6). In the case of samples subjected to additional heat treatment, the greatest differentiation of the mid-life hysteresis loops surface areas was observed during the tests at the highest value of the total strain amplitude ε_{ac} = 0.50%.

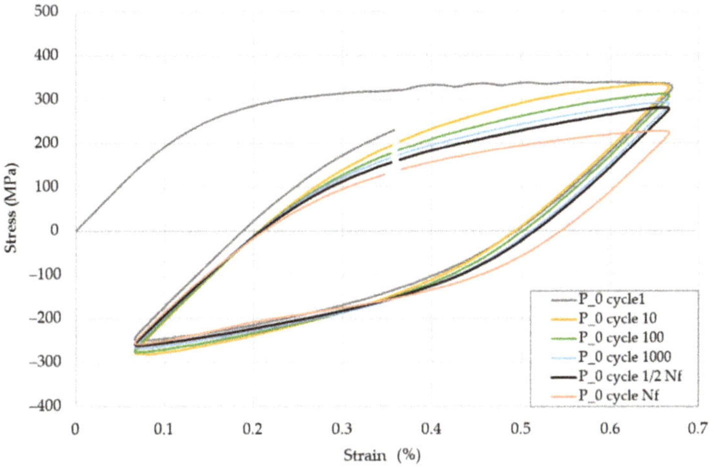

Figure 5. Hysteresis loops for selected cycles for ε_{ac} = 0.3% for characteristic cycles during low cycle fatigue (LCF) testing.

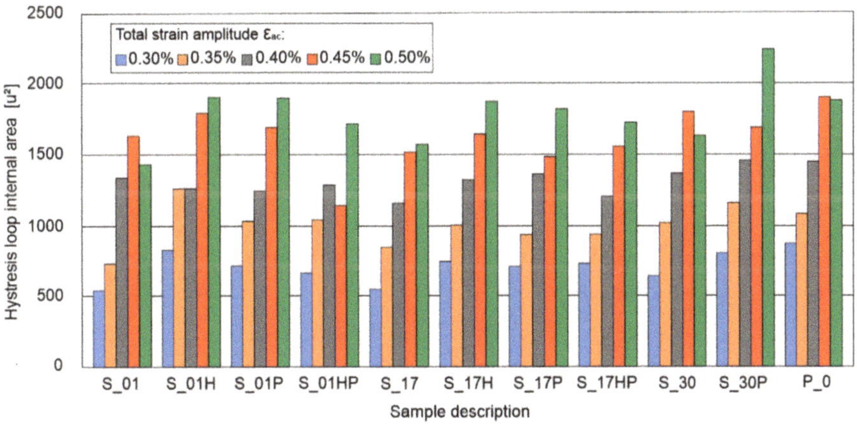

Figure 6. Registered areas of mid-life hysteresis loops (in square units).

The biggest hysteresis loop area, higher by 18.5% than the P_0 samples loop area, was found during the S_30P samples tests. Other samples tested at ε_{ac} = 0.50% were characterized by area range from −8.9% to 1.3% compared to the results obtained for the reference P_0 sample. In the case of the remaining samples, the observed differences in the hysteresis loop areas differed by 10–12% from the results obtained for the reference P_0 sample.

Stabilized hysteresis loop analysis allowed to determine the amplitudes of stress and plastic strain. Stress amplitude versus plastic strain amplitude can be described by the following equation:

$$log\sigma_a = logK' + n'log\varepsilon_{ap} \quad (2)$$

where:

σ_a—fatigue strength coefficient (MPa);
ε_{ap}—fatigue ductility coefficient;
K'—cyclic strength coefficient;
n'—cyclic strain hardening exponent.

Calculation results were presented in a chart with a log–log scale in Figure 7.

As it is depicted in Figure 7, almost all AM samples are characterized by higher cyclic stress response at a given strain level, except S_17P samples. The phenomenon of the increased value of this parameter is connected with a higher yield strength of as-built AM parts [15]. Registered higher values of cyclic stress response in AM parts are connected with the LCF testing characteristic, where the strain level was a constant value during the test. Significantly increased yield strength of AM samples gave an increase in cyclic stress response during LCF testing. Heat treatment of AM parts decreased cyclic stress response and made it similar to conventionally made material, and it S_17P this value is even smaller than in conventionally made samples. This phenomenon could be connected with an increased value of porosity in those samples after precipitation heat treatment. Further analysis of S_17P samples' fatigue behavior was taken into account in microfractures analysis.

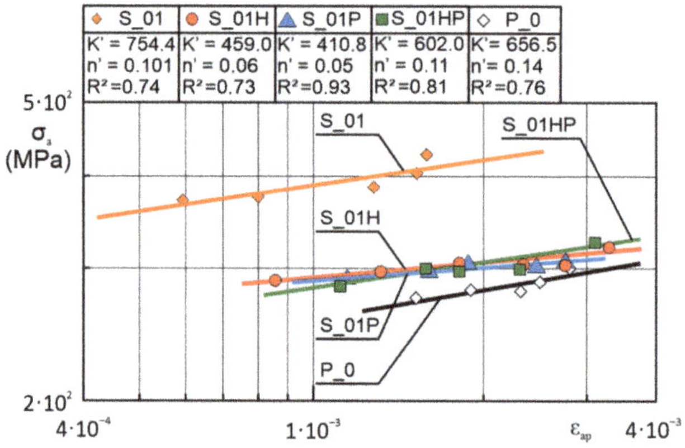

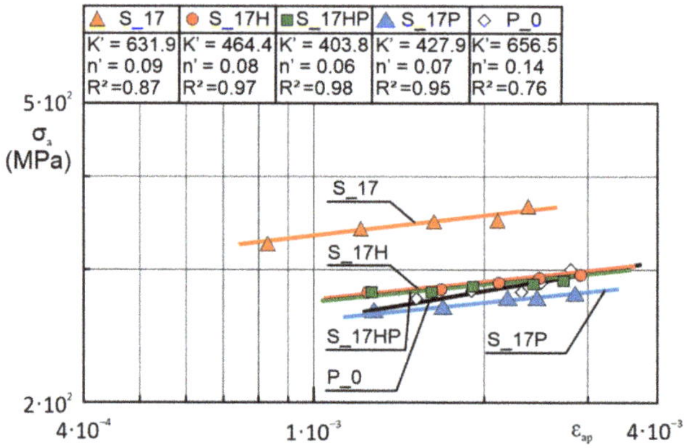

Figure 7. *Cont.*

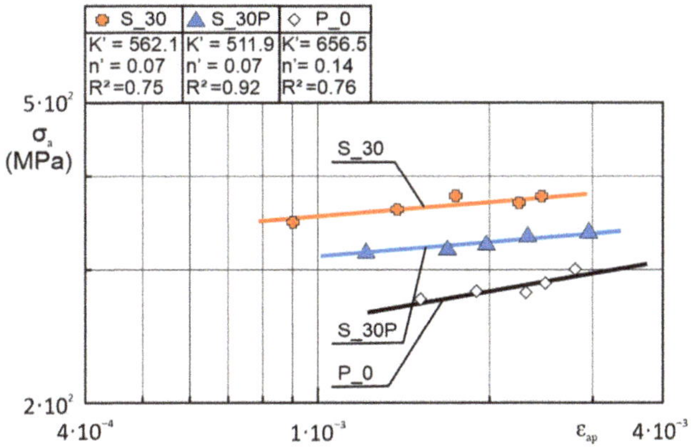

Figure 7. Stress amplitude versus plastic strain amplitude of as-built, heat-treated AM parts compared to conventionally made material stabilized hysteresis loops in log–log coordinates.

Equation (2) for each sample's group has the following form:

$$log\sigma_{a\ S_{01}} = log754.4 + 0.101log(\varepsilon_{ap});$$

$$log\sigma_{a\ S_{01H}} = log459.0 + 0.06log(\varepsilon_{ap});$$

$$log\sigma_{a\ S_{01P}} = log410.8 + 0.05log(\varepsilon_{ap});$$

$$log\sigma_{a\ S_{01HP}} = log602.0 + 0.11log(\varepsilon_{ap});$$

$$log\sigma_{a\ S_{17}} = log631.7 + 0.09log(\varepsilon_{ap});$$

$$log\sigma_{a\ S_{17H}} = log631.7 + 0.09log(\varepsilon_{ap});$$

$$log\sigma_{a\ S_{17HP}} = log427.9 + 0.07log(\varepsilon_{ap});$$

$$log\sigma_{a\ S_{17P}} = log410.8 + 0.05log(\varepsilon_{ap});$$

$$log\sigma_{a\ S_{30}} = log562.1 + 0.07log(\varepsilon_{ap});$$

$$log\sigma_{a\ S_{30P}} = log511.9 + 0.07log(\varepsilon_{ap});$$

$$log\sigma_{a\ P_{0}} = log656.5 + 0.14log(\varepsilon_{ap})$$

Based on the recorded data, further fatigue analysis was followed on the Morrow equation:

$$\varepsilon_{ac} = \varepsilon_{ae} + \varepsilon_{ap} = \frac{\sigma'_f}{E}(2N_f)^b + \varepsilon'_f(2N_f)^c \quad (3)$$

where:

ε_{ac}—total strain amplitude;
ε_{ae}—elastic strain amplitude; ε_{ap}—plastic strain amplitude;
E—Young's modulus;
σ'_f—fatigue strength coefficient;
ε'_f—fatigue ductility coefficient;
b—fatigue strength exponent;
c—fatigue ductility exponent.

Obtained results (Figure 8) allow stating that heat-treated additive manufactured 316L steel had a significant share of the plastic component during the destruction process; however, in the case of most samples, this share was significantly lower than in the case of non-heat-treated samples. It was particularly visible in the graphs concerning the samples S_17H (Figure 8f), S_17P (Figure 8g) and S_17HP (Figure 8h), where the angles between the lines corresponding to the amplitude of the plastic component ε_{ap} and the amplitude of the elastic component ε_{ae} were much smaller than in the case of the as-built sample S_17 (Figure 8e).

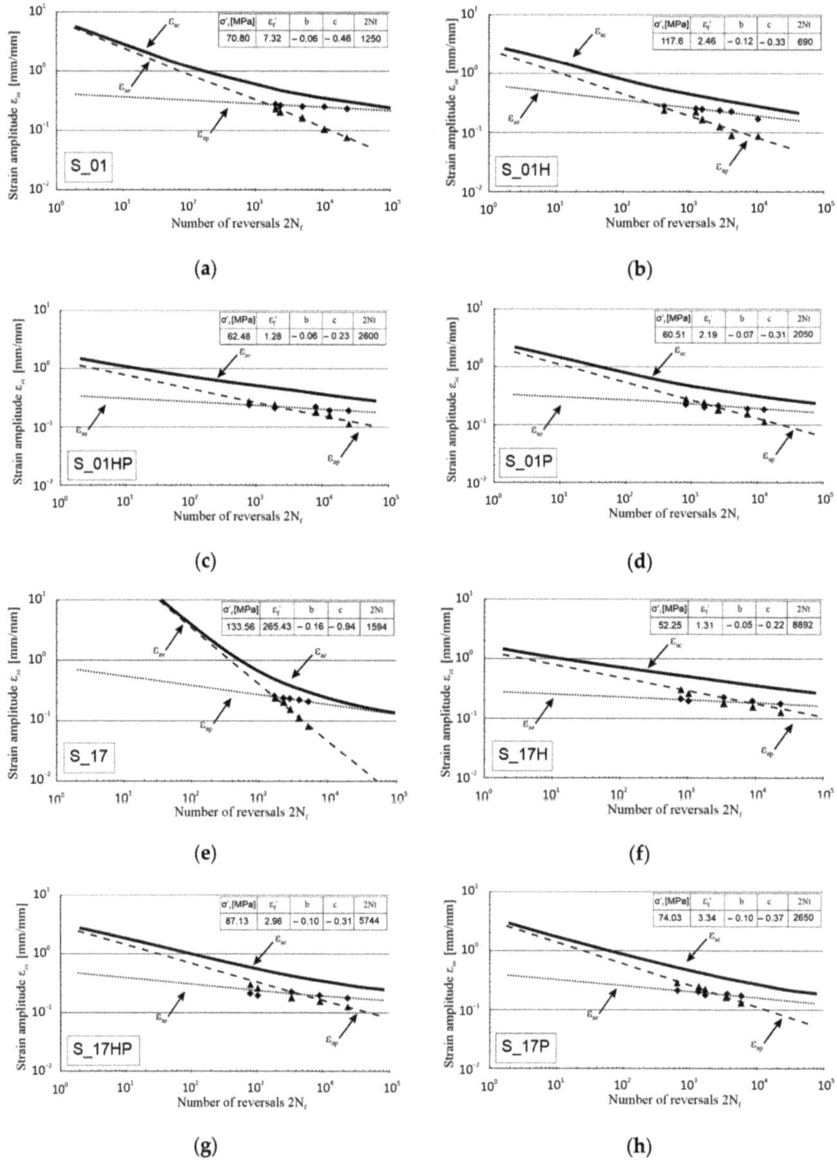

Figure 8. *Cont.*

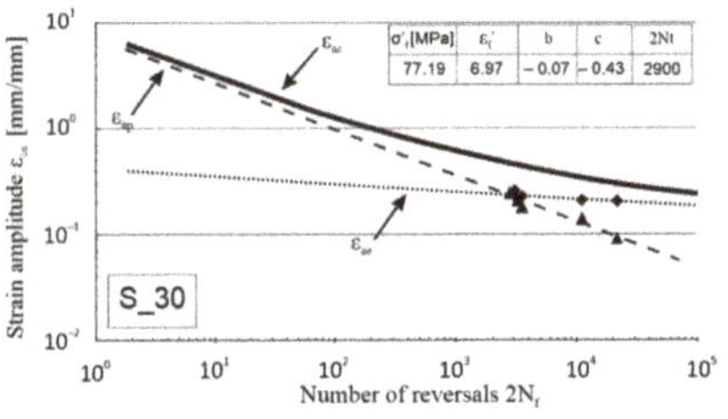

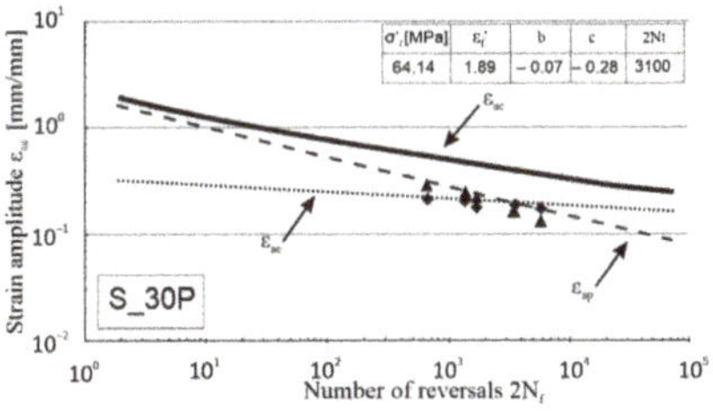

(i) **(j)**

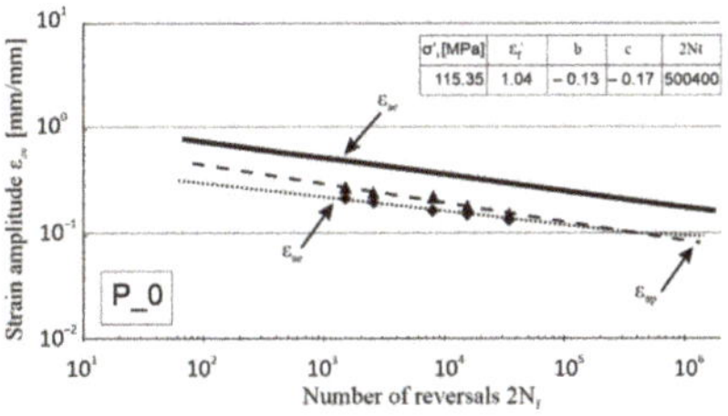

(k)

Figure 8. Number of half-cycle reversals vs. strain amplitude in log–log coordinates for as-built samples, heat-treated and conventionally made material ((**a**)—S_01, (**b**)—S_01H, (**c**)—S_01HP, (**d**)—S_01P, (**e**)—S_17, (**f**)—S_17H, (**g**)—S_17HP, (**h**)—S_17P, (**i**)—S_30, (**j**)—S_30P, (**k**)—P_0).

Conventionally produced 316L steel (P_0 samples) was characterized by the lowest share of the plastic component in total deformation of tested samples in comparison to additively manufactured parts (Figure 8k).

3.2. Fatigue Fractures Analysis

To better understand material fatigue behavior after a different type of treatment– microstructures from the author's previous research [15] of samples in each condition (as-built, before and after heat treatment) are shown in Figure 9.

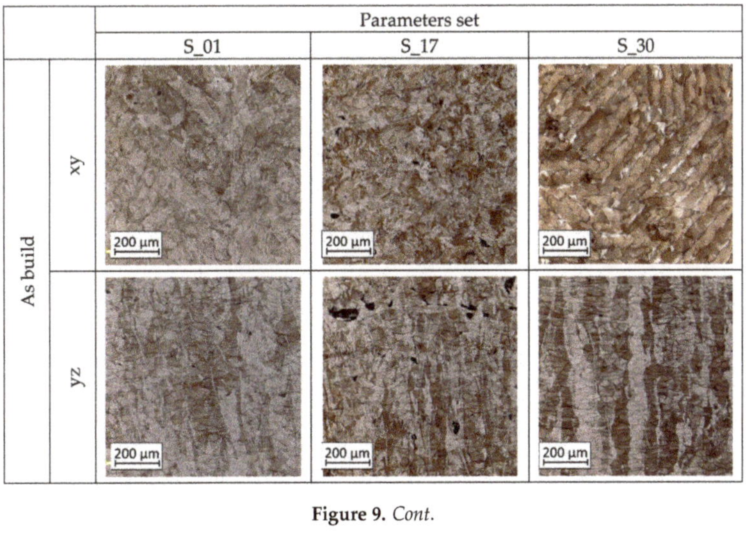

Figure 9. *Cont.*

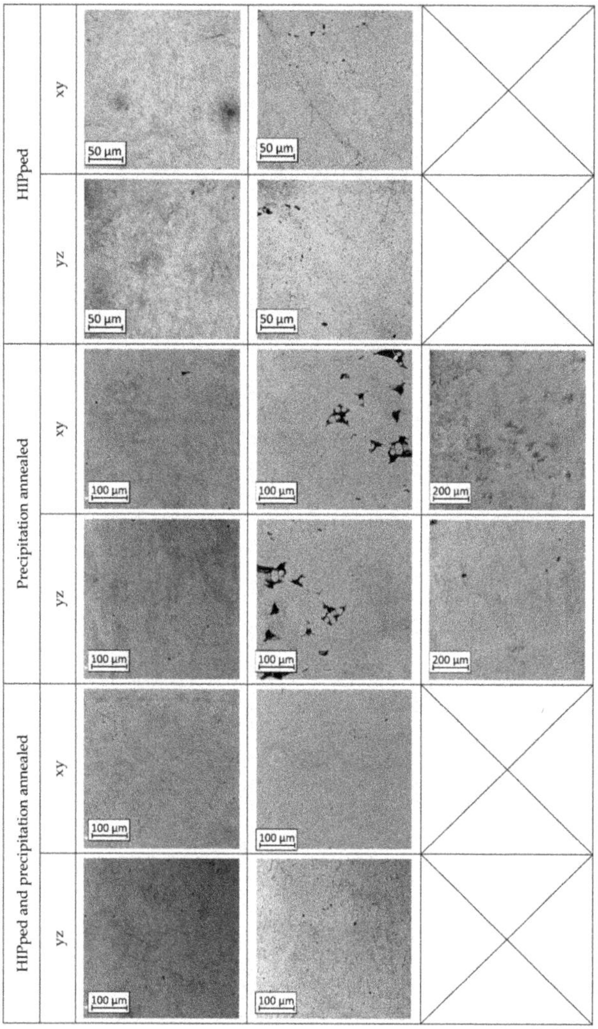

Figure 9. The microstructure of S_01, S_17 and S_30 samples heat-treated using different processes and their combinations [15].

Fractographic observations were made for all parameter groups (S_01, S_17, and S_30) and all heat-treated combinations of samples with additional analysis of conventionally made material. Fracture images of samples tested under two conditions: ε_{ac} = 0.30% and ε_{ac} = 0.50% are shown, respectively. The column "a" shows the entire fracture area and the crack propagation direction, marked using white arrows. In column "b", there are selected areas of the enlarged fractures.

Based on the fracture's observations (Figure 10), additively manufactured parts before and after heat treatment were characterized by plastic fractures with fatigue striations, as well as conventionally made P_0 samples. It is worth emphasizing that the morphology of the fracture surfaces of S_01H, S_17H, S_01HP, and S_17HP (all after heat treatment) samples was less complex, which also applied to the reference sample P_0. HIP and precipitation heat-treatment after the previous HIP eliminated the layered structure of additively manufactured parts, which significantly reduced the occurrence of multiplanar cracking. A noticeably greater number of multiplanar cracking was visible in samples subjected to precipitation heat treatment (S_01P, S_17P, S_30P). The significantly different cracking character of those samples may be related to the share of voids in the material structure. Fatigue fractures of the S_17P samples were characterized by a high number of voids and unmelted grains compared to other samples subjected to precipitation heat treatment (S_01P and S_30P). This phenomenon was strictly connected with the high initial porosity of as-built S_17 samples.

The observed structural heterogeneity was caused by the presence of random various voids sizes and share of unmelted grains, which have significantly different properties after heat treatment in comparison to the remaining volume of the material. This phenomenon affects the occurrence of local concentrations of residual stresses in the material structure. Mentioned imperfections significantly affect the development of microcracks inside the material in many parallel planes. Additionally, in S_17P samples, there were registered areas of cracks connection which cause occurring of the local breakage cracks. A significant increase in the share of voids in the S_17P samples completely changed the nature of fatigue cracking of this material. Most of the cracking sources occurred inside the material volume, especially at the most complex geometry of a specific void.

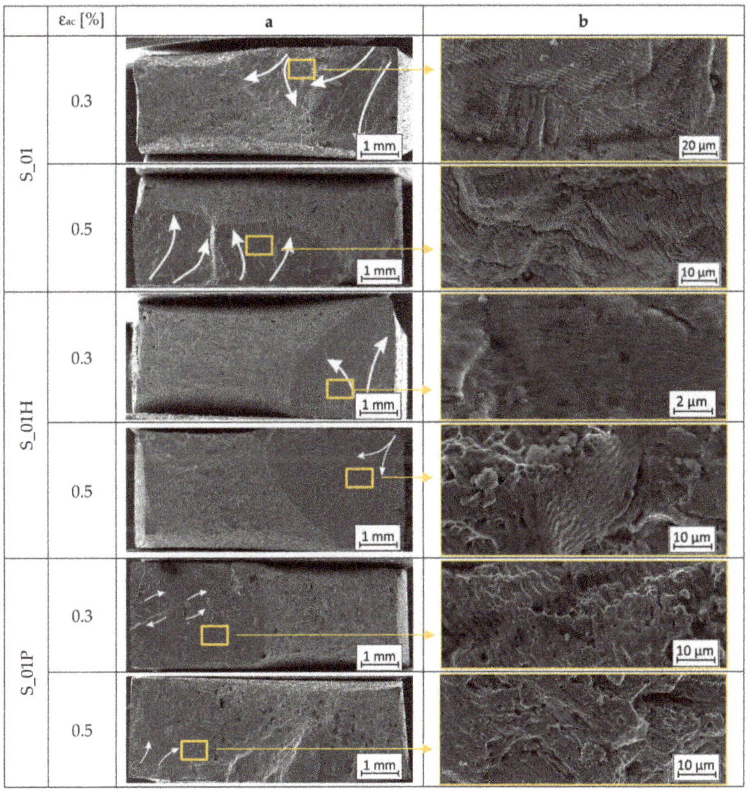

Figure 10. *Cont.*

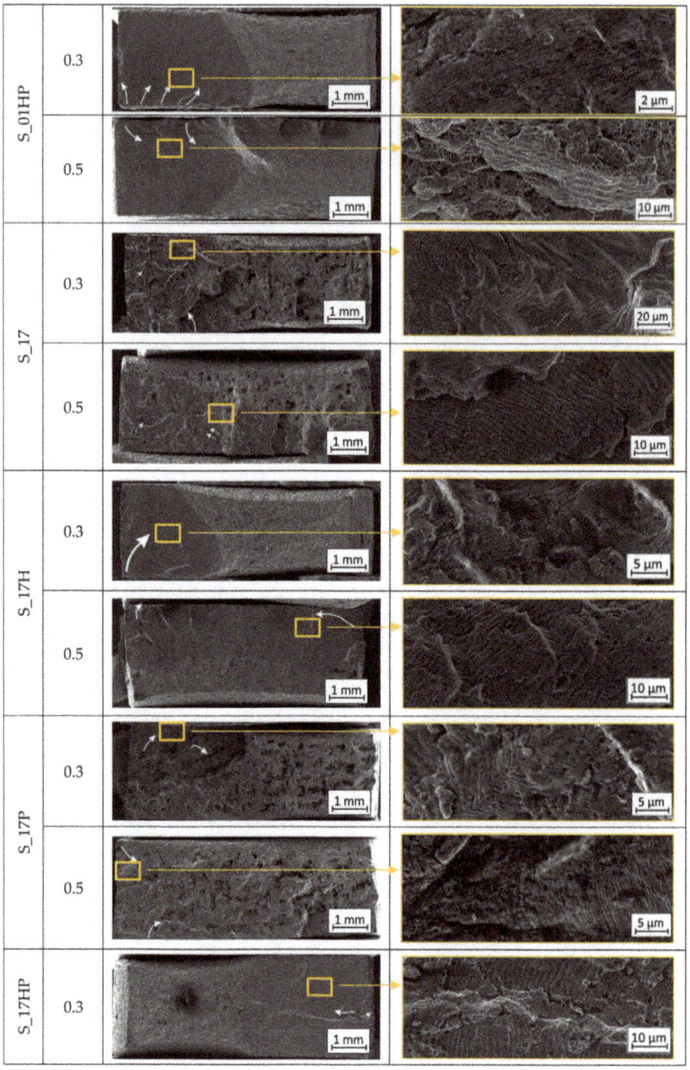

Figure 10. *Cont.*

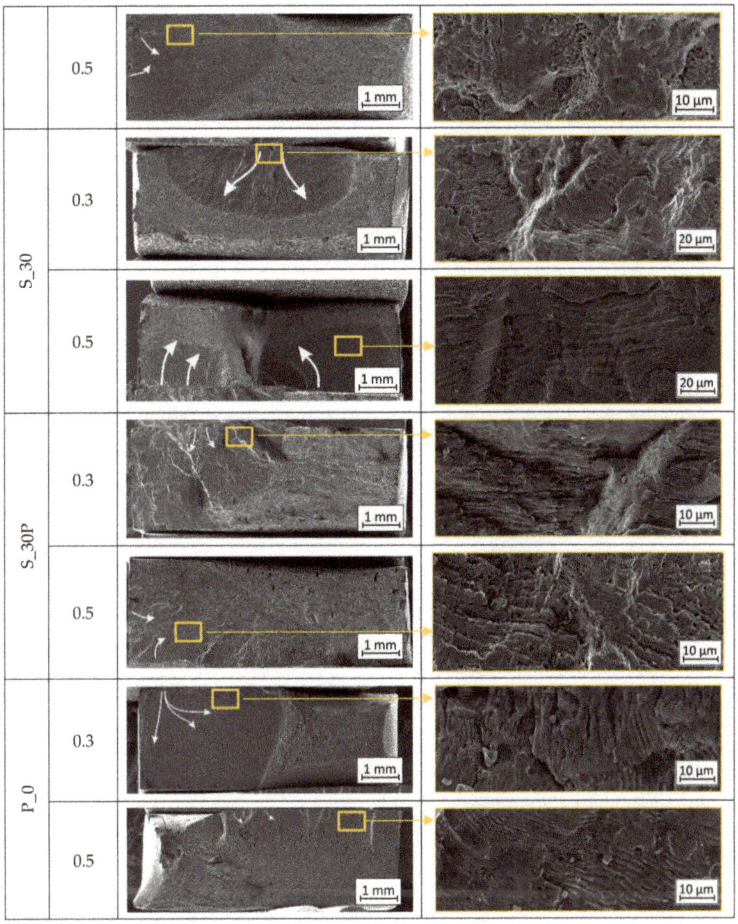

Figure 10. Fractures of low-cycle tested samples ((**a**)—macrostructure image, (**b**)—microfracture with visible fatigue striations).

4. Conclusions

The performed research of low cycle fatigue of SLM-processed and heat-treated parts allowed drawing the following conclusions:

1. Fatigue life charts created based on the Morrow equation exposed that the strain courses of additively manufactured parts were characterized by a significant share of the plastic component in the process of sample fractures above $\varepsilon_{ac} = 0.45\%$;
2. Conventionally made 316L steel (P0) required a lower stress level than additively produced samples, caused by the greater ductility of the conventionally produced material in comparison to AM samples;
3. Microfractures analysis showed that the additively manufactured samples were characterized by a more complex morphology than conventionally produced parts. The observed layered structure of as-built samples affected the occurrence of microcracks and multiplanar cracking;
4. Additively manufactured material, without any heat treatment, weakened throughout the entire load cycle range. At the same time—conventionally made material (and also additively

manufactured after heat treatment) were characterized by a visible stabilization at some cycles range;

5. The values of stress amplitude σ_a versus the number of load reversals showed much smaller changes of stress amplitude in the entire range of load cycle number N than in the case of non-heat-treated samples (less tendency to weakening of the material). In samples S_01H, S_01P, and S_01HP, the cyclic weakening stage is transient;
6. The obtained strain courses of additively manufactured parts, including samples subjected to additional heat treatment, indicated a significant share of a plastic component in the process of sample fractures.

Author Contributions: Conceptualization. L.Ś., K.G. and J.K.; methodology. K.G., A.O. and K.P.; software. J.T. and I.S.; formal analysis. L.Ś.; investigation. J.K., K.P. and I.S.; data curation. J.K. and M.W.; writing—original draft preparation. J.K. and M.M.; writing—review and editing. J.T. and M.M.; visualization. M.W. and J.K.; supervision. L.Ś.; project administration. L.Ś. and K.G.; funding acquisition. K.G. and A.O. All authors have read and agreed to the published version of the manuscript.

Funding: The research was supported by the university research grant No. 22-758.

Conflicts of Interest: The authors declare no conflict of interest.

References

1. Mooney, B.; Kourousis, K.I. A review of factors a ff ecting the mechanical a review of of maraging factors affecting properties of powder bed fusion. *Metals* **2020**, *10*, 1273. [CrossRef]
2. Gu, D.; Chen, H. Selective laser melting of high strength and toughness stainless steel parts: The roles of laser hatch style and part placement strategy. *Mater. Sci. Eng.* **2018**, *725*, 419–427. [CrossRef]
3. Jia, D.; Li, F.; Zhang, Y. 3D-Printing process design of lattice compressor impeller based on residual stress and deformation. *Sci. Rep.* **2020**, *10*, 1–11. [CrossRef] [PubMed]
4. Rokicki, P.; Kozik, B.; Budzik, G.; Dziubek, T.; Bernaczek, J.; Przeszlowski, L.; Markowska, O.; Sobolewski, B.; Rzucidlo, A. Manufacturing of aircraft engine transmission gear with SLS (DMLS) method. *Aircr. Eng. Aerosp. Technol.* **2016**, *88*, 397–403. [CrossRef]
5. Hopkinson, N.; Dickens, P. Rapid prototyping for direct manufacture. *Rapid Prototyp. J.* **2001**, *7*, 197–202. [CrossRef]
6. Malachowski, J.; Damaziak, K.; Platek, P.; Sarzynski, M.; Kupidura, P.; Wozniak, R.; Zahor, M. Numerical and experimental failure analysis of rifle extractor. *Eng. Fail. Anal.* **2016**, *62*, 112–127. [CrossRef]
7. Platek, P.; Damaziak, K.; Malachowski, J.; Kupidura, P.; Wozniak, R.; Zahor, M. Numerical study of modular 5.56 mm standard assault rifle referring to dynamic characteristics. *Def. Sci. J.* **2015**, *65*, 431–437. [CrossRef]
8. Kucewicz, M.; Baranowski, P.; Małachowski, J.; Popławski, A.; Płatek, P. Modelling, and characterization of 3D printed cellular structures. *Mater. Des.* **2018**, *142*, 177–189. [CrossRef]
9. Antolak-Dudka, A.; Płatek, P.; Durejko, T.; Baranowski, P.; Małachowski, J.; Sarzyński, M.; Czujko, T. Static and dynamic loading behavior of Ti6Al4V honeycomb structures manufactured by Laser Engineered Net Shaping (LENSTM) technology. *Materials* **2019**, *12*, 1225. [CrossRef]
10. Kucewicz, M.; Baranowski, P.; Stankiewicz, M.; Konarzewski, M.; Płatek, P.; Małachowski, J. Modelling and testing of 3D printed cellular structures under quasi-static and dynamic conditions. *Thin Walled Struct.* **2019**, *145*, 106385. [CrossRef]
11. Sienkiewicz, J.; Płatek, P.; Jiang, F.; Sun, X.; Rusinek, A. Investigations on the mechanical response of gradient lattice structures manufactured via SLM. *Metals* **2020**, *10*, 213. [CrossRef]
12. Śniezek, L.; Grzelak, K.; Torzewski, J.; Kluczyński, J. Study of the mechanical properties components made by SLM additive technology. In Proceedings of the 11th International Conference on Intelligent Technologies in Logistics and Mechatronics Systems (ITELMS), Panevezys, Lithuania, 28–29 April 2016; pp. 145–153.
13. Kluczyński, J.; Śnieżek, L.; Kravcov, A.; Grzelak, K.; Svoboda, P.; Szachogłuchowicz, I.; Franek, O.; Morozov, N.; Torzewski, J.; Kubeček, P. The examination of restrained joints created in the process of multi-material FFF additive manufacturing technology. *Materials* **2020**, *13*, 903. [CrossRef] [PubMed]

14. Kluczyński, J.; Śniezek, L.; Grzelak, K.; Mierzyński, J. The influence of exposure energy density on porosity and microhardness of the SLM additive manufactured elements. *Materials* **2018**, *11*, 2304. [CrossRef]
15. Kluczyński, J.; Śniezek, L.; Grzelak, K.; Oziebło, A.; Perkowski, K.; Torzewski, J.; Szachogłuchowicz, I.; Gocman, K.; Wachowski, M.; Kania, B. Comparison of different heat treatment processes of selective laser melted 316L steel based on analysis of mechanical properties. *Materials* **2020**, *13*, 3805. [CrossRef] [PubMed]
16. Kluczyński, J.; Sniezek, L.; Grzelak, K.; Torzewski, J. The influence of layer re-melting on tensile and fatigue strength of selective laser melted 316L steel. In Proceedings of the 12th International Conference on Intelligent Technologies in Logistics and Mechatronics Systems (ITELMS), Panevezys, Lithuania, 26–27 April 2018; pp. 115–123.
17. Yadollahi, A.; Shamsaei, N.; Thompson, S.M.; Elwany, A.; Bian, L. Effects of building orientation and heat treatment on fatigue behavior of selective laser melted 17-4 PH stainless steel. *Int. J. Fatigue* **2017**, *94*, 218–235. [CrossRef]
18. Lewandowski, J.J.; Seifi, M. Metal additive manufacturing: A review of mechanical properties. *Annu. Rev. Mater. Res.* **2016**, *46*, 151–186. [CrossRef]
19. Leuders, S.; Lieneke, T.; Lammers, S.; Tröster, T.; Niendorf, T. On the fatigue properties of metals manufactured by selective laser melting—The role of ductility. *J. Mater. Res.* **2014**, *29*, 1911–1919. [CrossRef]
20. Zhang, M.; Sun, C.N.; Zhang, X.; Wei, J.; Hardacre, D.; Li, H. High cycle fatigue and ratcheting interaction of laser powder bed fusion stainless steel 316L: Fracture behaviour and stress-based modelling. *Int. J. Fatigue* **2019**, *121*, 252–264. [CrossRef]
21. Riemer, A.; Richard, H.A. Crack propagation in additive manufactured materials and structures. *Procedia Struct. Integr.* **2016**, *2*, 1229–1236. [CrossRef]
22. Kluczyński, J.; Śniezek, L.; Grzelak, K.; Torzewski, J.; Szachogłuchowicz, I.; Wachowski, M.; Łuszczek, J. Crack growth behavior of additively manufactured 316L steel-influence of build orientation and heat treatment. *Materials* **2020**, *13*, 3259. [CrossRef]
23. Nezhadfar, P.D.; Burford, E.; Anderson-Wedge, K.; Zhang, B.; Shao, S.; Daniewicz, S.R.; Shamsaei, N. Fatigue crack growth behavior of additively manufactured 17-4 PH stainless steel: Effects of build orientation and microstructure. *Int. J. Fatigue* **2019**, *123*, 168–179. [CrossRef]
24. Blinn, B.; Ley, M.; Buschhorn, N.; Teutsch, R.; Beck, T. Investigation of the anisotropic fatigue behavior of additively manufactured structures made of AISI 316L with short-time procedures PhyBaL LIT and PhyBaL CHT. *Int. J. Fatigue* **2019**, *124*, 389–399. [CrossRef]
25. Awd, M.; Tenkamp, J.; Hirtler, M.; Siddique, S.; Bambach, M.; Walther, F. Comparison of microstructure and mechanical properties of Scalmalloy® produced by selective laser melting and laser metal deposition. *Materials* **2017**, *11*, 17. [CrossRef] [PubMed]
26. Romano, S.; Patriarca, L.; Foletti, S.; Beretta, S. LCF behaviour and a comprehensive life prediction model for AlSi10Mg obtained by SLM. *Int. J. Fatigue* **2018**, *117*, 47–62. [CrossRef]
27. Bressan, S.; Ogawa, F.; Itoh, T.; Berto, F. Low cycle fatigue behavior of additively manufactured Ti-6Al-4V under non-proportional and proportional loading. *Frat. Integrità Strutt.* **2019**, *13*, 18–25. [CrossRef]
28. Tucho, W.M.; Lysne, V.H.; Austbø, H.; Sjolyst-Kverneland, A.; Hansen, V. Investigation of effects of process parameters on microstructure and hardness of SLM manufactured SS316L. *J. Alloy. Compd.* **2018**, *740*, 910–925. [CrossRef]
29. Andreau, O.; Pessard, E.; Koutiri, I.; Penot, J.D.; Dupuy, C.; Saintier, N.; Peyre, P. A competition between the contour and hatching zones on the high cycle fatigue behaviour of a 316L stainless steel: Analyzed using X-ray computed tomography. *Mater. Sci. Eng.* **2019**, *757*, 146–159. [CrossRef]
30. Shifeng, W.; Shuai, L.; Qingsong, W.; Yan, C.; Sheng, Z.; Yusheng, S. Effect of molten pool boundaries on the mechanical properties of selective laser melting parts. *J. Mater. Process. Technol.* **2014**, *214*, 2660–2667. [CrossRef]
31. Elangeswaran, C.; Cutolo, A.; Muralidharan, G.K.; de Formanoir, C.; Berto, F.; Vanmeensel, K.; Van Hooreweder, B. Effect of post-treatments on the fatigue behaviour of 316L stainless steel manufactured by laser powder bed fusion. *Int. J. Fatigue* **2019**, *123*, 31–39. [CrossRef]
32. Horovistiz, A.L.; de Campos, K.A.; Shibata, S.; Prado, C.C.S.; Hein, L.R. de O. Fractal characterization of brittle fracture in ceramics under mode I stress loading. *Mater. Sci. Eng.* **2010**, *527*, 48474850. [CrossRef]

33. Cabanettes, F.; Joubert, A.; Chardon, G.; Dumas, V.; Rech, J.; Grosjean, C.; Dimkovski, Z. Topography of as built surfaces generated in metal additive manufacturing: A multi scale analysis from form to roughness. *Precis. Eng.* **2018**, *52*, 249–265. [CrossRef]
34. Kluczyński, J.; Śnieżek, L.; Grzelak, K.; Janiszewski, J.; Płatek, P.; Torzewski, J.; Szachogłuchowicz, I.; Gocman, K. Influence of selective laser melting technological parameters on the mechanical properties of additively manufactured elements using 316L austenitic steel. *Materials* **2020**, *13*, 1449. [CrossRef] [PubMed]
35. Wang, D.; Liu, Y.; Yang, Y.; Xiao, D. Theoretical and experimental study on surface roughness of 316L stainless steel metal parts obtained through selective laser melting. *Rapid Prototyp. J.* **2016**, *22*, 706–716. [CrossRef]
36. Lo, K.H.; Shek, C.H.; Lai, J.K.L. Recent developments in stainless steels. *Mater. Sci. Eng. Rep.* **2009**, *65*, 39–104. [CrossRef]
37. *ASTM E466 96. Standard Practice for Conducting Force Controlled Constant Amplitude Axial Fatigue Tests of Metallic Materials*; ASTM Headquarters: West Conshohocken, PA, USA.
38. *ASTM Description E 606-80. Standard Recommended Practice for Constant-Amplitude Low-Cycle Fatigue Testing*; ASTM Headquarters: West Conshohocken, PA, USA.
39. *PN-84/H-04334 Low Cycle Fatigue Testing of Metallic Materials*; Standards Committee of Poland (PKN): Warsaw, Poland.
40. Ho, H.S.; Zhou, W.L.; Li, Y.; Liu, K.K.; Zhang, E. Low-Cycle fatigue behavior of austenitic stainless steels with gradient structured surface layer. *Int. J. Fatigue* **2020**, *134*, 105481. [CrossRef]

materials MDPI

Article

Possibilities of the Utilization of Ferritic Nitrocarburizing on Case-Hardening Steels

Jiri Prochazka [1], Zdenek Pokorny [1,*], Jozef Jasenak [2], Jozef Majerik [2] and Vlastimil Neumann [3]

1 Department of Mechanical Engineering, Faculty of Military Technology, University of Defence, 662 10 Brno, Czech Republic; jiri.prochazka@unob.cz
2 Department of Manufacturing Technologies and Materials, Faculty of Special Technology, Alexander Dubcek University of Trencin, 911 06 Trencin, Slovakia; jozef.jasenak@tnuni.sk (J.J.); jozef.majerik@tnuni.sk (J.M.)
3 Department of Combat and Special Vehicles, Faculty of Military Technology, University of Defence, 662 10 Brno, Czech Republic; vlastimil.neumann@unob.cz
* Correspondence: zdenek.pokorny@unob.cz; Tel.: +420-973-442-839

Abstract: This paper is devoted to the possibilities of the utilization of chosen chemical heat treatment technologies on steels used for manufacturing highly stressed components of military vehicles and weapons systems. The technologies chosen for this research are plasma ferritic nitrocarburizing and ferritic nitrocarburizing in a gaseous atmosphere. These technologies were applied on a steel equivalent 1.5752 (i.e., CSN 41 6426), which is suitable for carburizing. Chemical composition of the steel was verified by optical emission spectrometry. An observation of a microstructure and an assessment of the parameters of obtained white layers were performed by optical microscopy. Morphology and porosity of the surface were observed by electron microscopy. The depth of diffusion layers was evaluated in accordance with ISO 18203:2016(E) from the results of microhardness measurements. A friction coefficient was obtained as a result of measurements in accordance with a linearly reciprocating ball-on-flat sliding wear method. Wear resistance was assessed by employing the scratch test method and a profilometry. The profilometry was also utilized for surface roughness assessment. It was proved that both tested chemical heat treatment technologies are suitable for surface treatment of the selected steel. Both technologies, ferritic nitrocarburizing in plasma and a gaseous atmosphere, are beneficial for the improvement of surface properties and could lead to a suppression of geometrical deformation in comparison with frequently utilized carburizing. Moreover, the paper presents a procedure that creates a white layer-less ferritic nitrocarburized surface by utilizing an appropriate modification of chemical heat treatment parameters, thus subsequent machining is no longer required.

Keywords: chemical heat treatment; ferritic nitrocarburizing; friction coefficient; wear resistance; microhardness profile; depth of diffusion layer; white layer thickness; case-hardening steel

Citation: Prochazka, J.; Pokorny, Z.; Jasenak, J.; Majerik, J.; Neumann, V. Possibilities of the Utilization of Ferritic Nitrocarburizing on Case-Hardening Steels. *Materials* **2021**, *14*, 3714. https://doi.org/10.3390/ma14133714

Academic Editors: Marcin Wachowski, Henryk Paul and Sebastian Mróz

Received: 7 June 2021
Accepted: 30 June 2021
Published: 2 July 2021

1. Introduction

The goal of the paper is to clarify the possibilities of modifying the surface properties of highly stressed components of military vehicles and weapons systems. In the case of these components, including gears, camshafts, pins of crankshafts, connecting rods, parts of weapon systems, etc., high resistivity of the surface against abrasion and corrosion is required. These components are also exposed to combined mechanical stress. Due to these facts, the high hardness and corrosion resistance of the surface is required contrary to the core, which must stay tough enough to provide banding and impact resistance [1]. The requirement of mutually exclusive properties like hardness and toughness is a motivation for utilizing surface heat and chemical heat treatments, coating, or bimetal cladding, which allows for a combination of mentioned properties simultaneously [2–4]. The product of these technologies is a surface with properties which differ from an original structure [5].

Surface heat treatments and chemical heat treatments are utilized most often in manufacturing of the mentioned exposed parts. These technologies include surface hardening, carburizing, carbonitriding, nitriding, ferritic nitrocarburizing, boriding, etc. [6–8]. Despite the many disadvantages of carburizing, this technology is utilized very frequently. Carburizing is based on a carbon diffusion from the atmosphere, composed of a solid (charcoal powder), liquid, or gaseous medium saturated by carbon, into the surface of a steel. A gaseous atmosphere is used the most often. The disadvantage of carburizing is a high temperature between 800–1000 °C, with the need for subsequent hardening of carburized products [9]. Cracks and geometrical deformations may occur due to the severe decrease in temperature when hardening. The low accuracy of product geometry after carburizing, required for subsequent grinding, is a motivation to find an alternative technology [10]. This paper is devoted to the possibilities of utilizing one such substitutional technology under consideration, namely ferritic nitrocarburizing [11,12]. Contrary to carburizing, the main advantage of ferritic nitrocarburizing is the lower temperature of exposition, generally between 537–600 °C [5]. In the field of thermochemical diffusion techniques, only nitrocarburizing, as well as nitriding, does not require quenching as a subsequent procedure. It results in higher dimension precision without the need for subsequent machining.

2. Materials and Methods

One type of steel, carefully selected by an analysis of materials used in the manufacturing of mentioned parts, was investigated by chemical composition verification, microstructure and surface morphology observation, microhardness measurement, friction coefficient measurement and wear resistance assessment.

2.1. Materials

According to an analysis of materials used for manufacturing highly stressed components of military vehicles and weapons systems, the steel equivalent 1.5752 (i.e., CSN 41 6426), which is utilized for the manufacturing of mechanical gears, crankshafts, connecting rods, etc., is suitable for carburizing and was therefore selected.

2.2. Chemical Composition

The conformity of the chemical composition of the steel was verified by using the advanced CCD optical emission spectrometer Tasman Q4 (Bruker, Billerica, MA, USA), utilizing the Fe110 method. Results are obtained as an average value of five measurements.

2.3. Specimen Preparation

Four disk-shaped specimens were cut off from the steel rod in its normalized state. Heat treatment was performed subsequently. Heat-treated steel was determined as an initial state for further chemical heat treatment processes. All specimens were heat-treated in accordance with the parameters shown in Table 1.

Table 1. Parameters of heat treatment.

Parameter	Hardening	Tempering
Temperature (°C)	870	600
Time (min)	20	60
Medium	Water	Water

Datasheets of the selected case-hardening steel primarily contain parameters of a heat treatment which follows carburizing. Thus, in this case where the carburizing was not performed, the heat-treatment parameters appropriate for commonly utilized low-carbon alloyed steels were chosen in accordance with the literature and experiences of the department.

The heat treatment was followed by grinding and finished by using sandpaper F-1000 according to FEPA. Three of the specimens prepared in this manner were subsequently chemical heat treated.

The first specimen, later used as a reference, was left in a heat-treated state. The second specimen was ferritic nitrocarburized by employing a gaseous atmosphere in a NITREX appliance. The ferritic nitrocarburizing chamber was tempered to 530 °C and the exposure time was set to 6 h. The third specimen was plasma ferritic nitrocarburized. The temperature and process duration were the same as in the previous case.

A cross section of each specimen was cut off by a metallographic saw and molded into thermoplastic powder. Preparation of specimens' surfaces was completed by grinding and polishing. Polishing was performed by using a velvet and diamond paste with grains of size 0.5 μm.

After a performance of all experimental measurements described in the following text, based on the results, the fourth specimen obtained by plasma ferritic nitrocarburizing, with the same atmospheric parameters but a shortened process duration, was treated and subsequently subjected to the experimental measurements.

2.4. Microstructure

First, a microstructure of the specimen in its heat-treated state was observed by using an Olympus DSX500i optical microscope (Olympus, Tokyo, Japan). Due to the reason of migration of carbon and nitrogen into the specimen´s surface, and due to the creation of a specific surface layer during the ferritic nitrocarburizing process, an area influenced by chemical heat treatment was also observed when using the microscope.

A nitride layer composed of three sublayers is formed during ferritic nitrocarburizing. The creation of such a layer is also typical for nitriding. On the top of the surface, the first sublayer, called the compound or white layer, composed of nitrides and carbonitrides of iron, is mostly created. The second sublayer, a diffusion layer, is formed by a dispersion of carbides, nitrides, and carbonitrides of iron and alloying elements. The last sublayer is known as a transition area, which is situated between the diffusion layer and the core microstructure [13–15].

The process of ferritic nitrocarburizing is always accompanied by the formation of the white layer as a part of the nitride layer. After supersaturation of this sublayer by carbon and nitrogen, defined interstitial elements diffuse further into the surface [13]. Although the white layer provides protection against corrosion and wear, and improves initial sliding properties after assembly, in some conditions it can negatively affect surface properties [9]. Porosity of the white layer is the most unfavorable factor. Pores decrease corrosion resistance and primarily increase the brittleness of the white layer created on the top of the surface. It may cause the initiation of cracks due to mechanical stress [16]. The porosity of the surface of the chemically heat treated specimens was observed by the TESCAN MIRA 4th generation scanning electron microscope, with a magnification of 10,000×.

2.5. Microhardness Measurement

The depth of the diffusion layer was measured in agreement with ISO 18203:2016(E) [17]. NHD (nitriding hardness depth) was evaluated by using microhardness profiles obtained as the result of microhardness measurements conducted by automated microhardness tester LM247 AT LECO (Leco Corporation, St. Joseph, MI, USA). Microhardness profiles were measured on cross sections of specimens. The pattern of impressions was 0.8 mm long. The first impression was performed at a distance of 20 μm from the surface. The step of following impressions was set to 10 μm. The load of the Vickers indenter was 100 g. The resulting microhardness profiles were established from mean values of three measurements of the microhardness profiles, which were performed on each specimen. The depth of the diffusion layer was determined as distance from the surface, where the microhardness of the material´s core increased by 50 HV was measured [17]. Microhardness of the core was determined as a mean value from three impressions, measured at a sufficient distance from a surface, rounded to the nearest multiple of 10 HV [17].

2.6. Surface Roughness Measurement

Roughness is deviation from the ideal shape of the surface. Besides machining processes, a chemical heat treatment also plays a role in modifying the surface microgeometry [18]. During application of selected chemical heat treatments, the surface is subjected to high-temperature-enhanced diffusion of elements already present in the material, as well as elements gathered from the surrounding atmosphere which condensate on the surface. In plasma applications, sputtering also affects the surface microgeometry [18]. Due to that fact, surface roughness is also involved in producing values of the friction coefficient. The measurement of roughness profile parameters, i.e., Ra, Rq, Rt, Rz and RSm performed on the profilometer Talysurf CLI1000 (Taylor Hobson Ltd, Leicester, England) by utilizing the contact method was also included in the experimental method. According to ISO 4288, the first measurement, expecting the roughest profile, was performed to obtain values of the Ra and Rz parameters. Those were subsequently utilized for the selection of lr, roughness sampling length (also known as cut-off), and ln, roughness evaluation length, as parameters for further measurements [19]. Values of the roughness profile parameters were finally determined as average values from ten profiles of each measurement.

2.7. Measurement of Coefficient of Friction

The friction coefficient, defined as a ratio of friction force and a normal force by which an indenter is loaded, is a parameter which describes the efficiency of a contact movement [12]. Thus, the friction coefficient measurement performed according to the ball-on-flat method, described by the standard ASTM G133-05, was incorporated into the experimental method [20]. The method utilizes ongoing monitoring of parameters, such as normal and friction forces, friction coefficient, acoustic emission, etc., during the measurement, while the indenter moves linearly reciprocally along the specimen surface where the tribological wear track is formed. The Bruker UMT-3 TriboLab instrument (Bruker, Billerica, MA, USA), corresponding with conditions defined by the standard ASTM G133-05, was utilized for the measurement.

The main parameter of such a performed measurement is the friction coefficient, which is plotted to the graph dependent on time. Due to an uneven reciprocating motion of the indenter, obtained data have a square waves profile and hence average values are obtained by subsequent software filtering [21]. Generally, it is possible to distinguish running-in and steady-state wear from the graphs. During the running-in phase the contact surfaces adapt and polish each other [22]. An uneven friction coefficient value is recorded during this phase. When the value stabilize itself, the steady-state wear phase is reached. This period typically takes a long time and in most cases the measurement ends in this phase.

2.8. Wear Resistance Assessment

A lower friction coefficient does not always mean better wear resistance. The opposite has already been described, and ferritic nitrocarburized as well as nitrided surfaces are not an exception [23]. For this reason, subsequent measurements were performed on the same instrument by utilizing the scratch test method described in the ASTM G171-03 and ASTM G1624-05 standards [24,25]. The method utilizes Rockwell´s HRC indenter instead of ball-on-flat which was used in the previous case. During each measurement, solely one linear scratch of a predetermined length is performed on the specimen surface. Two modifications of the load of the indenter, such as constant and linearly increasing load, are allowed by the standards. The first measurement was performed with a linearly increasing load in a range from 0 N to 50 N. During the measurement, at a certain distance from an initial point of the measurement some failures of the surface occur. According to the ASTM G1624-05, the first ruptures which occur in the scratch are marked as location L_{C1}, which characterize a normal force causing cohesive failures in the white layer. A location where the failures lead to spalling of the white layer is marked L_{C2}. According to the result of the first measurement, three values (15 N, 20 N and 35 N) were selected for further measurements performed by an indenter with constant load.

3. Results

The beneficial influence of selected ferritic nitrocarburizing on surface properties of steel equivalent 1.5752 (i.e., CSN 41 6426) was experimentally assessed according to chemical composition, microstructure, microhardness, thickness and porousness of the white layer, kinetic friction coefficient, surface roughness and wear resistance.

3.1. Chemical Composition

Results of the measurement of chemical composition and the limits of the content of chemical elements mentioned in a datasheet of the steel are listed in Table 2.

Table 2. Chemical composition of steel equivalent 1.5752 (i.e., CSN 41 6426); instrument: Tasman Q4 Bruker (wt.%).

C	Mn	Si	Cr	Ni	P	S
			OES/Bulk			
0.12	0.49	0.33	0.76	3.20	0.020	0.006
			Datasheet			
0.10–0.17	0.30–0.60	0.17–0.37	0.60–0.90	2.70–3.20	<0.035	<0.035

The results of the measurement concur with the values listed in the material datasheet of steel equivalent 1.5752 (i.e., CSN 41 6426). Chemical composition has an influence not only on properties of the core of the material but also on the properties of surface layers obtained by chemical heat treatment. According to the literature [9,26], knowledge of the content of elements allows us to predict properties of the resulting layers.

In the case of steel equivalent 1.5752 (i.e., CSN 41 6426), a low content of carbon predetermines the steel for chemical heat treatment, such as carburizing and ferritic nitrocarburizing. By these processes, an increase in the content of interstitial elements, like carbon and nitrogen, in a surface layer is supported. Higher content of carbon and nitrogen in cooperation with alloying elements such as chromium, vanadium and molybdenum leads to the precipitation of particularly hard carbides and nitrides during the diffusion process in the surface layer [26]. Therefore, it can be assumed that the content of chromium in steels may cause an increase in the microhardness of the surface layer after nitrocarburizing. From the measurement, it follows that the steel contains a relatively high amount of nickel. This substitutional element negatively influences the creation of carbides and nitrides as well as the diffusion of carbon and nitrogen into the material [26]. The depth of the nitride layer also depends on the concentration of nitrogen [27]. By utilizing the concentration of nickel, about 3.2 wt.%, it is possible to predict the decrease in the depth of the nitride layer.

3.2. Microstructure

The microstructure of heat-treated steel, etched by 2% NITAL, which was observed on a cross-sectional specimen by an Olympus DSX500i, is shown in Figure 1. The observed microstructure corresponds with the parameters of the heat treatment applied to the specimens.

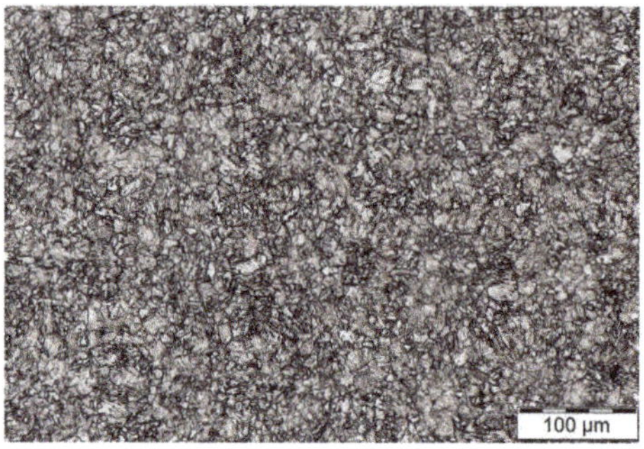

Figure 1. Microstructure of heat-treated specimen. Magnification, 500×.

Chemically heat treated areas obtained by ferritic nitrocarburizing which were observed on the cross-sections of the specimens are shown in Figures 2 and 3. On the left side of Figure 2, the white layer obtained by 6 h of gaseous ferritic nitrocarburizing is shown. The thickness of the white layer oscillates around 21 µm. The porosity of the white layer extends to approximately half of its thickness. The porous part of the white layer is predominantly formed by condensation of particles from the gaseous atmosphere [28]. Due to its properties, mentioned in Section 2.4, it is suitable to remove this porous part of the white layer. This recommendation is unique and dependent on the manner of utilization.

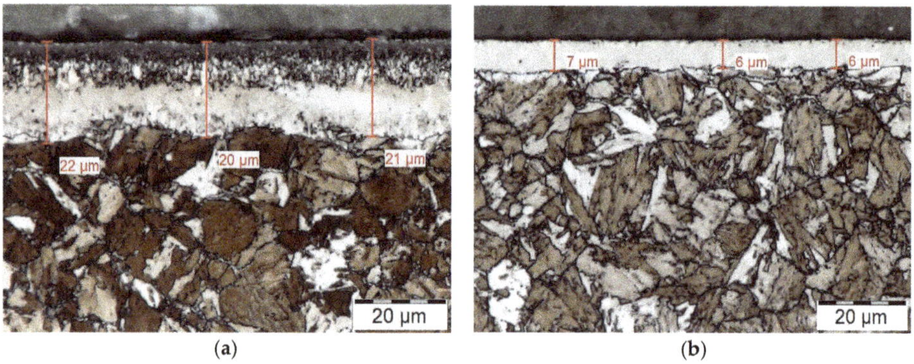

(**a**) (**b**)

Figure 2. White layers of chemically heat treated specimens. (**a**) Six hours of gaseous ferritic nitrocarburizing and (**b**) 6 h of plasma ferritic nitrocarburizing. Magnification, 2000×.

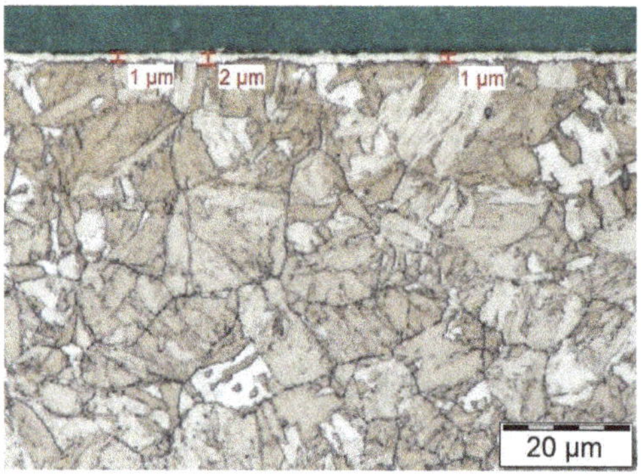

Figure 3. White layer after 4 h of gaseous ferritic nitrocarburizing. Magnification, 2000×.

The white layer of the specimen after plasma ferritic nitrocarburizing is shown on the right side of Figure 2. The white layer obtained by plasma ferritic nitrocarburizing is about three times thinner and almost pore-less in comparison with the previous case. Although properties of the white layer obtained after 6 h are better in this case, the paper aims to achieve a white layer-less surface to reduce the possible need of subsequent machining. Therefore, the process of plasma ferritic nitrocarburizing, applied subsequently to the fourth specimen, was shortened to a period of 4 h. An obtained white layer of this nature is visible in Figure 3.

As is shown in the Figure 3, by reducing the process duration a significant narrowing of the white-layer thickness was achieved. Thus, subsequent machining of the surface is no longer necessary. The statements of the previous paragraphs are further supported by pictures of the surface's morphology shown in Figure 4, obtained by a TESCAN MIRA 4th generation scanning electron microscope with a magnification of 10,000×.

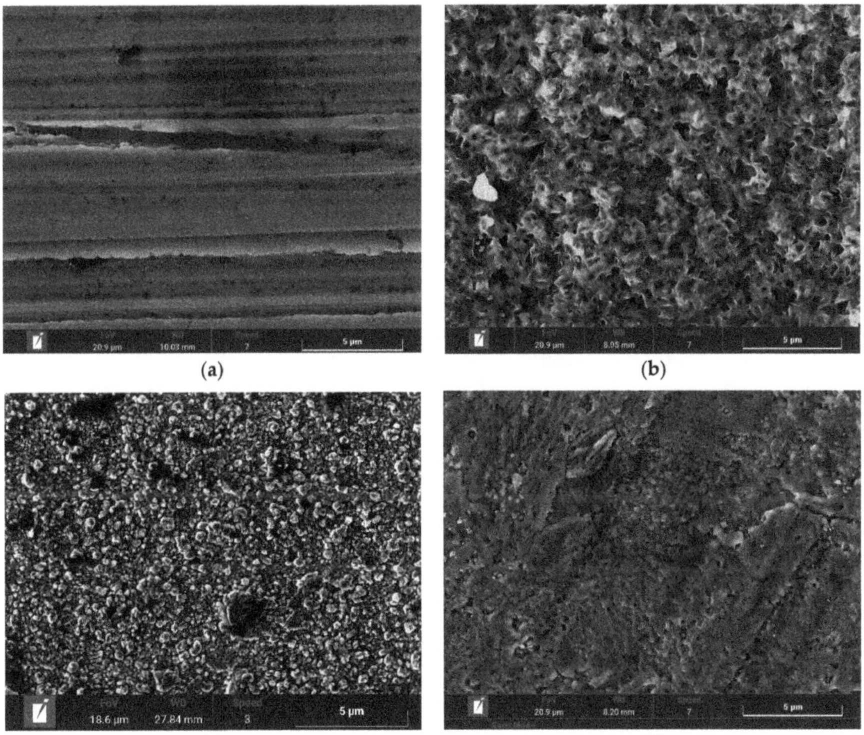

Figure 4. Surface morphology of specimens. (**a**) Heat-treated; (**b**) after 6 h of gaseous FNC; (**c**) after 6 h of plasma FNC; and (**d**) after 4 h of plasma FNC. Magnification, 10,000×.

The surface morphology of the plasma ferritic nitrocarburized specimens is more homogenous, with lower porosity in comparison with those nitrocarburized in the gaseous atmosphere. This finding is in accordance with statements published in [28].

3.3. Microhardness

As was mentioned in Section 2.5, the measurement of microhardness was utilized for an evaluation of the depth of the diffusion layers. Microhardness profiles composed with appropriate microstructures in the background are shown in Figures 5 and 6.

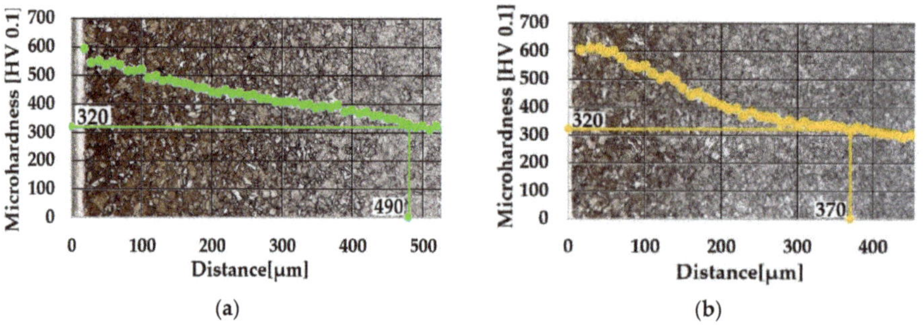

Figure 5. Profiles of microhardness. (**a**) Six hours of gaseous ferritic nitrocarburizing and (**b**) 6 h of plasma ferritic nitrocarburizing. Magnification of microstructure, 500×.

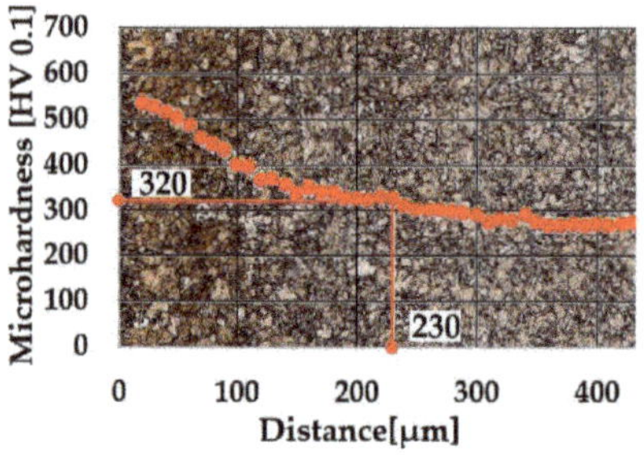

Figure 6. Microhardness profile of layer obtained by 4 h of plasma ferritic nitrocarburizing. Magnification of microstructure, 500×.

Microhardness of the core as 270 HV 0.1 was determined; therefore, the limit value of microhardness was equal to 320 HV. As is shown in Figure 5, the depth of the surface layer, obtained by 6 h of gaseous ferritic nitrocarburizing, was determined as NHD 320 HV 0.1 = 490 μm. In the case of 6 h of plasma ferritic nitrocarburizing, the depth of the layer achieved NHD 320 HV 0.1 = 370 μm. While differences in maximum values of microhardness are barely visible, a significant decrease of the NHD in cases of plasma ferritic nitrocarburizing in comparison with the process utilizing gaseous atmosphere was found. This phenomenon could be attributed to the different character of both processes [11]. As is visible in Figure 6, the phenomenon is enhanced with a shortening of the process duration.

Whereas the maximum values of microhardness were preserved in range of 550–600 HV in all cases of chosen chemical heat treatment processes, 4-hour plasma ferritic nitrocarburizing provided lower case depth, i.e., NHD 320 HV 0.1 = 230 μm in comparison with 6-h variants.

3.4. *Coefficient of Friction and Surface Roughness Measurement*

The friction coefficient measurement was performed subsequently. The measurement parameters were set as follows: normal force = 10 N, duration = 1000 s, frequency = 5 Hz, stroke length = 10 mm, temperature = 23 °C, diameter of tungsten carbide ball indenter = 6.35 mm and dry friction. The results of all measurements are plotted together into the graph listed in Figure 7.

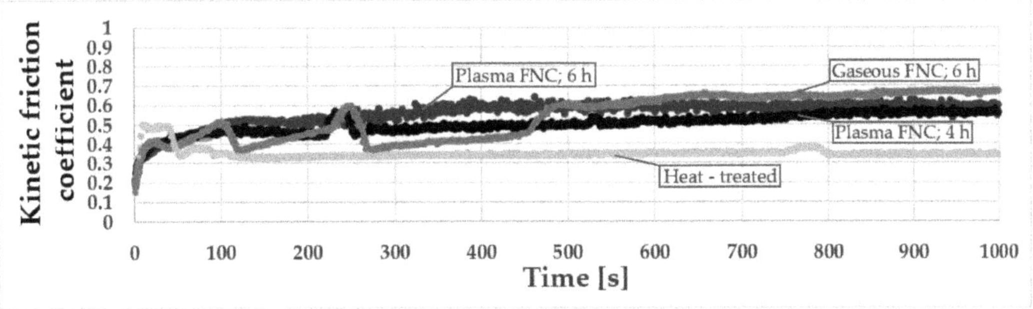

Figure 7. Friction coefficient comparison of heat-treated and ferritic nitrocarburized surfaces.

The Figure 7 implies that application of all selected chemical heat treatment processes caused a significant deterioration of surface friction coefficient in comparison with a solely heat-treated surface. To have a better chance of describing the phenomenon, the friction coefficient measurement was accompanied by the measurement of the roughness profile parameters listed in Table 3.

Table 3. The values of roughness profile parameters of surfaces subjected to the selected chemical heat treatments.

Parameter		Ground Surface	Gaseous Ferritic Nitrocarburizing; 6 h	Plasma Ferritic Nitrocarburizing; 6 h	Plasma Ferritic Nitrocarburizing; 4 h
Ra	(μm)	0.0705	0.2200	0.1020	0.0671
Rq	(μm)	0.1110	0.2770	0.1430	0.1090
Rt	(μm)	1.4300	1.9600	1.6200	1.5300
Rz	(μm)	1.3400	1.7600	1.4600	1.3900
RSm	(mm)	0.0130	0.0207	0.0161	0.0146

Note: measurement parameters were selected as follows: lr = 0.8 mm, ln = 4 mm.

By combining the information mentioned previously, it is possible to state that an increase in the surface microhardness made the surfaces slide by the indenter less easily due to a decrease in the surface formability and also the presence of hard C- and N-based debris in the wear track. The values of the friction coefficient in steady-state wear phase are sorted by order of the values of roughness profile parameters. Thus, the dependance of friction coefficient on surface roughness was confirmed. In the case of the friction coefficient of the gaseous ferritic nitrocarburized surface, the unevenness lasting for a period of almost 500 s could be caused by a warping of the porous part of the white layer during the measurement.

3.5. Wear Resistance Assessment

All specimens were subjected to three scratch test measurements at three different constant loads, i.e., 15 N, 20 N and 35 N. Images of such obtained tracks taken by an Olympus DSX500i opto-digital microscope are visible in Figure 8.

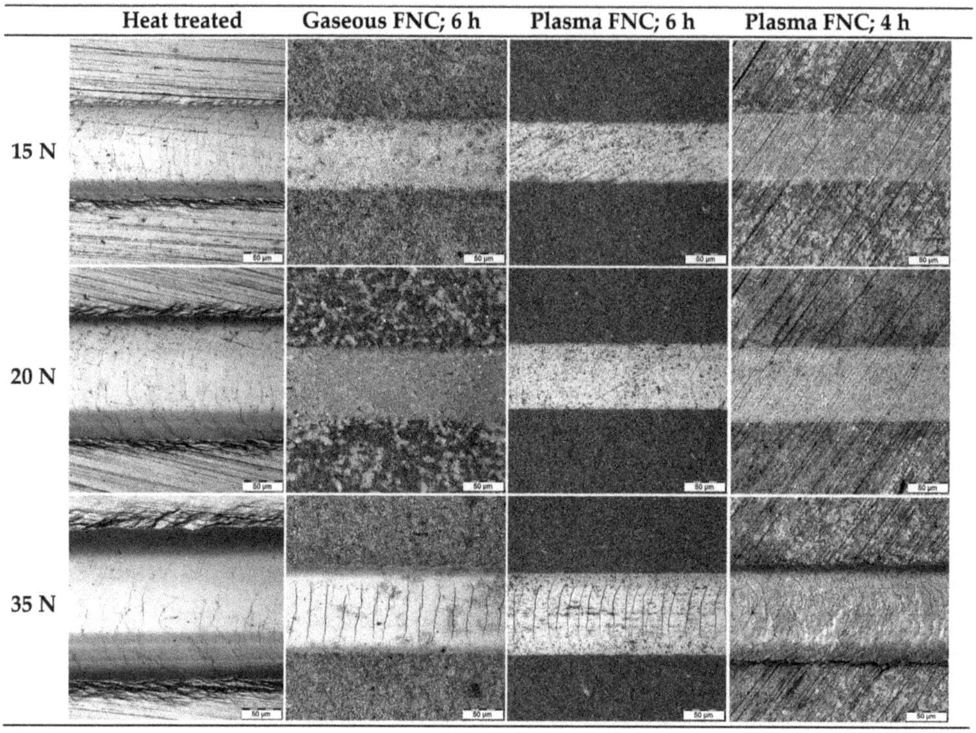

Note: The measurements were led in direction from left to right side

Figure 8. Images of the scratch test wear tracks. Magnification: 1000×.

In the case of the heat-treated specimen, similar cohesive failures occurred independently of the load. It could be caused by a higher plasticity of a softer surface in comparison with chemically heat-treated specimens. Due to a lower resistance of the heat-treated surface to being penetrated by the indenter, with increased load much wider tracks were obtained.

In both cases of plasma ferritic nitrocarburizing minor cohesive failures under load 20 N as well as at 35 N were visible. Although, in the case of shortened plasma ferritic nitrocarburizing an observable increase of track width occurred, but in neither case did spalling of the white layer appear.

Whereas significant cohesive failures were only visible at load 35 N in the case of gaseous ferritic nitrocarburizing, unexpected spalling of the white layer in an area of the track´s edge already occurred at 15 N. The phenomenon was documented by scanning electron microscopy and is visible in Figure 9.

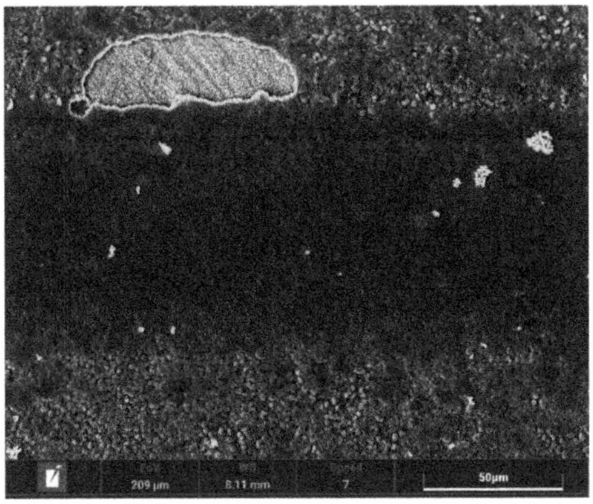

Figure 9. Spalling of the white layer; gaseous ferritic nitrocarburized surface; and load = 15 N. Magnification, 1000×.

From each scratch wear track, three cross-sectional profiles by a Talysurf CLI1000 profilometer were subsequently taken utilizing the contact method. Parameters of the cross-sectional profiles, such as track width, maximum depth and area of the profile were collected in Table 4 and subsequently utilized for the assessment of surface wear resistance.

Table 4. Values of cross-sectional profile parameters.

Parameter	Load	Heat Treated	Gaseous FNC	Plasma FNC; 6 h	Plasma FNC; 4 h
Width (μm)	15 N	81.67	80.50	62.70	63.80
	20 N	93.50	82.83	70.17	74.77
	35 N	123.00	95.73	83.80	91.60
Depth (μm)	15 N	2.62	1.15	0.64	0.87
	20 N	3.77	1.43	1.01	1.12
	35 N	8.43	2.62	2.11	2.84
Area (μm^2)	15 N	156.67	58.23	22.43	30.67
	20 N	256.33	82.13	42.77	54.37
	35 N	765.33	186.00	132.67	186.33

The results listed above imply that the application of the selected chemical heat treatments caused a significant decrease in track dimensions. Based on this fact, it could be stated that surface wear resistance was influenced beneficially by the application of the mentioned treatments. From this point of view, the best results were reached in the case of 6-hour plasma ferritic nitrocarburization. The impact of the shortened plasma process on the wear resistance seems to also be positive, and in many cases comparable values of scratch dimensions with ferritic nitrocarburizing performed in a gaseous atmosphere were measured.

4. Discussion

The possibility of utilizing selected ferritic nitrocarburizing technologies as a substitutional technology for chemical-heat treatment of case-hardening steel equivalent 1.5752 (i.e., CSN 41 6426), instead of carburizing, was tested. The low content of carbon (0.12 wt.%) predetermine the steel not only for carburizing but also for all surface treatments based on saturation of the surface microstructure by interstitial elements soluble in the ferritic or austenitic lattice. Such chemical heat treatments also include low-temperature treatments, such as nitriding or ferritic nitrocarburizing, which was tested in this paper.

Both technologies, ferritic nitrocarburizing in gaseous atmosphere and ferritic nitrocarburizing in plasma, were applied on the material in a heat-treated state. At first, a 6-hour process duration was selected for both technologies. In both cases, the microstructure in an area near the surface was affected and the layer composed of the white and diffusion layer was observed. Different characteristics of the technologies also caused significant differences in the nitrocarburized layers, especially in the white layers. Whereas in the case of plasma ferritic nitrocarburizing a pore-less white layer was observed with a 6 μm thickness, in the case of gaseous ferritic nitrocarburizing pores appeared in the outer half of its 20 μm thick white layer. The observed differences influenced the choice of a treatment technology of shortened duration for application on the last specimen. Four-hour plasma ferritic nitrocarburizing, subsequently selected to obtain a white layer-less surface, led to the creation of a solely 1 μm thick white layer.

The surface morphology was subsequently assessed by scanning electron microscopy. In the case of gaseous ferritic nitrocarburizing, a highly porous surface was observed. Existence of the porosity in the surface is a consequence of condensation of particles from the gaseous atmosphere, and thus is a consequence of the treatment technology. Therefore, the finding resulting from the observation of the white layer cross section was supported. In the case of plasma-utilizing technology the surface seemed to be smooth, almost pore-less and less rough in comparison with the ground surface of the solely heat-treated specimen. This phenomenon, that was even more visible in the case of shortened plasma ferritic nitrocarburizing, is possible to attribute to the character of the process. During plasma ferritic nitrocarburizing, ionized particles in the nitrocarburizing atmosphere accelerate and bombard the surface. Some surface particles are ejected from the surface under the bombardment, smoothening the surface during the process.

The microhardness measurements showed that a significant increase of the microhardness with maximum values in a range from 550 HV 0.1 to 600 HV 0.1 by all selected ferritic nitrocarburizing technologies was achieved. It could be attributed to an ability of chromium, also contained in the steel as 0.76 wt.% of it, to create hard and stable nitrides and carbides during the chemical heat treatment processes. By utilizing the microhardness profiles the nitriding hardness depth was determined. The deepest case of NHD, 320 HV 0.1 = 490 μm, was obtained in the gaseous atmosphere. Although in cases of the plasma technology the maximum values of the microhardness were similar, exponentially decreasing microhardness when approaching the core caused a narrowing of the diffusion layer in comparison with ferritic nitrocarburizing in the gaseous atmosphere, where the microhardness decreased linearly. Hence, the ferritic nitrocarburizing potential of the gaseous atmosphere seems to be a little greater in comparison with the selected plasma ferritic nitrocarburizing. The lowest NHD 320 HV 0.1 = 230 μm was found in the case of the shortened process of plasma ferritic nitrocarburizing. Thus, the assumption relating to the interstitial element diffusion retardation, caused by the great amount of nickel (3.2 wt.%) contained in the steel was confirmed, and in connection with the short process duration led to the creation of the shallow nitrocarburized layer.

In all cases of ferritic nitrocarburizing, a significant increase of the friction coefficient was visible. The graph of the kinetic friction coefficient reflects an increase in the surface microhardness as well as the surface roughness, which both contribute to the surface friction coefficient modification. The lowest values of the roughness profile parameters were measured on the shortened plasma ferritic nitrocarburized surface. The friction coefficient measured on the same surface, the lowest of the chemically heat-treated surfaces, could be affected solely by the low roughness of the surface. Contrarily, the highest friction coefficient appeared in the case of gaseous ferritic nitrocarburizing, where the highest values of roughness profile parameters were also found. It also complies with the observed surface morphology. Whereas in both cases of plasma ferritic nitrocarburizing smooth and fluent friction coefficient graphs were obtained, in the case of gaseous ferritic nitrocarburizing a significant fluctuation of the friction coefficient values during the first 500 s was visible. It could be caused by a warping of the porous part of the white layer.

The figures and the cross-sectional profiles of the scratches, obtained by the scratch test measurement at three different loads, 15 N, 20 N and 35 N, were utilized for the wear resistance assessment. The observation and measurements showed that significant increases in the surface wear resistance by application of the selected chemical heat treatments were reached. The best results were in the case of 6-hour plasma ferritic nitrocarburizing, where increasing load caused only a minor increase in the scratch dimensions and solely a cohesive failure in the scratch track. Similar results were almost observed in the case of the shortened plasma process, but the load of 35 N caused a greater increase of the track's dimensions in comparison with other chemical heat treatments. While in the cases of plasma ferritic nitrocarburized surfaces adhesive failures of the white layer at neither load appeared, 15 N was already enough for an occurrence of unexpected delamination in the white layer on the gaseous ferritic nitrocarburized surface. As well as in the case of the friction coefficient measurement, the phenomenon could be associated with an existence of the porous part of the white layer which is brittle and thus prone to be delaminated.

5. Conclusions

The information listed above is possible to be summarized into a statement: steel equivalent 1.5752 (i.e., CSN 41 6426) is appropriate for the selected ferritic nitrocarburizing technologies. In cases where deep surface layers are not demanded, plasma ferritic nitrocarburizing seem to be more appropriate for a reason, that reason being that thin and pore-less white layers, providing enhancement in corrosion resistance, can be obtained. Thus, in comparison with frequently utilized carburizing, for an economic reason, the plasma ferritic nitrocarburizing could be a more convenient surface treatment due to the absence of a need for subsequent machining and the shorter exposure time of this technology.

Author Contributions: Conceptualization, J.P.; methodology, Z.P. and J.P.; formal analysis, Z.P. and J.P.; investigation, J.P., Z.P., J.J., J.M. and V.N.; resources, J.P., Z.P., J.J., J.M. and V.N.; writing—original draft preparation, J.P. and Z.P.; writing—review and editing, Z.P. and J.P.; supervision, Z.P. All authors have read and agreed to the published version of the manuscript.

Funding: This research was funded by the specific research project 2020 "SV20-216" at the Department of Mechanical Engineering, University of Defence in Brno, the Project for the Development of the Organization "DZRO Military autonomous and robotic systems" and the Slovak Research and Development Agency under contract No. APVV-15-0710.

Institutional Review Board Statement: Not applicable.

Informed Consent Statement: Not applicable.

Data Availability Statement: Data are contained within the article.

Conflicts of Interest: The authors declare no conflict of interest.

References

1. Rauscher, J. *Vozidlové Motory: Studijní Opory*; VUT FSI: Brno, Czech Republic, 2004.
2. Panfil-Pryka, D.; Kulka, M.; Makuch, N.; Michalski, J.; Dziarski, P. The Effect of Temperature Distribution during Laser Heat Treatment of Gas-Nitrided 42CrMo4 Steel on the Microstructure and Mechanical Properties. *Coatings* **2020**, *10*, 824. [CrossRef]
3. Kosturek, R.; Wachowski, M.; Śnieżek, L.; Gloc, M. The Influence of the Post Weld Heat Treatment on the Microstructure of Inconel 625/Carbon Steel Bimetal Joint Obtained by Explosive Welding. *Metals* **2019**, *9*, 246. [CrossRef]
4. Kurek, A.; Wachowski, M.; Niesłony, A.; Płociński, T.; Kurzydłowski, K.J. Fatigue Tests and Metallographic of Explosively Cladded Steel-Titanium Bimetal/Badania Zmęczeniowe i Metalograficzne Bimetalu Stal-Tytan Zgrzewanego Wybuchowo. *Arch. Metall. Mater.* **2014**, *59*, 1566–1570. [CrossRef]
5. Totten, G.E. *Steel Heat Treatment: Metallurgy and Technologies*, 2nd ed.; Taylor & Francis: Boca Raton, FL, USA, 2007.
6. Callister, W.D. *Materials Science and Engineering: An Introduction*, 7th ed.; John Wiley & Sons: New York, NY, USA, 2007.
7. Dobrocky, D.; Studeny, Z.; Pokorny, Z.; Pospichal, M.; Smida, O. Effect of plasma nitriding on the notch toughness of spring steel. In Proceedings of the METAL 2016, 25th Anniversary International Conference on Metallurgy and Materials, Brno, Czech Republic, 25–27 May 2016; TANGER Ltd.: Ostrava, Czech Republic, 2016; pp. 1037–1044.

8. Studeny, Z.; Pokorny, Z.; Dobrocky, D. Fatigue properties of steel after plasma nitriding process. In Proceedings of the METAL 2016, 25th Anniversary International Conference on Metallurgy and Materials, Brno, Czech Republic, 25–27 May 2016; TANGER Ltd.: Ostrava, Czech Republic, 2016; pp. 1175–1180.
9. Hruby, V.; Tulka, J.; Kadlec, J. *Povrchové Technologie*; Vojenská akademie v Brně: Brno, Czech Republic, 1995.
10. Dybowski, K.; Sawicki, J.; Kula, P.; Januszewicz, B.; Atraszkiewicz, R.; Lipa, S. The Effect of the Quenching Method on the Deformations Size of Gear Wheels agter Vacuum Carburizing. *Arch. Metall. Mater.* **2016**, *61*, 1057–1062. [CrossRef]
11. Pye, D. *Practical Nitriding and Ferritic Nitrocarburizing*; ASM International: Phoenix, AZ, USA, 2003.
12. Davis, J.R. *Surface Engineering for Corrosion and Wear Resistance*; ASM International: Phoenix, AZ, USA, 2001.
13. Ratajski, J. Relation between phase composition of compound zone and growth kinetics of diffusion zone during nitriding of steel. *Surf. Coat. Technol.* **2009**, *203*, 2300–2306. [CrossRef]
14. Malinova, T.; Malinov, S.; Pantev, N. Simulation of microhardness profiles for nitrocarburized surface layers by artificial neural network. *Surf. Coat. Technol.* **2001**, *135*, 258–267. [CrossRef]
15. Hong, Y.; Wu, C.L.; Chen, J.H. Precipitation of γ' nitrides in N-saturated ferrite at high temperature and its effect on nitriding. *J. Alloys Compd.* **2019**, *792*, 818–827. [CrossRef]
16. Dobrocký, D.; Pokorný, Z.; Studený, Z.; Šurláková, M. Influence of the carbonitriding to change the surface topography of 16MnCr5 steel. In Proceedings of the METAL 2017, 26th Anniversary International Conference on Metallurgy and Materials, Brno, Czech Republic, 24–26 May 2017; TANGER Ltd.: Ostrava, Czech Republic, 2017; pp. 1085–1091.
17. ISO. *Steel—Determination of the Thickness of Surface-Hardened Layer*; ISO: Geneva, Switzerland, 2016; ISO 18203:2016(E).
18. Dobrocky, D.; Pokorny, Z.; Studeny, Z.; Faltejsek, P.; Dostal, P.; Polakova, N. Change of surface texture parameters of ground surfaces after application of hard and abrasion resistant layers. *ECS Trans.* **2018**, *87*, 431–442. [CrossRef]
19. ISO. *Geometric Product Specification (GPS)—Surface Texture—Profile Method: Rules and Procedures for the Assessment of Surface Texture*; ISO: Geneva, Switzerland, 1998; BS EN ISO 4288:1998.
20. ASTM International. *Standard Test Method for Linearly Reciprocating Ball-on-Flat Sliding Wear*; ASTM International: West Conshohocken, PA, USA, 2010; ASTM G133-05.
21. Bayer, R.G. *Mechanical Wear Fundamentals and Testing*; M. Dekker: New York, NY, USA, 2004.
22. Bhushan, B. *Modern Tribology Handbook*; CRC Press: Boca Raton, FL, USA, 2001.
23. Prochazka, J.; Pokorny, Z.; Dobrocky, D. Service behavior of nitride layers of steels for military applications. *Coatings* **2020**, *10*, 975. [CrossRef]
24. ASTM International. *Standard Test Method for Scratch Hardness of Materials Using a Diamond Stylus*; ASTM International: West Conshohocken, PA, USA, 2017; ASTM G171-03.
25. ASTM International. *Standard Test Method for Adhesion Strenght and Mechanical Failure Modes of Ceramic Coatings by Quantitative Single Point Scratch Testing*; ASTM International: West Conshohocken, PA, USA, 2010; ASTM C1624-05.
26. Pokorny, Z.; Dobrocky, D.; Kadlec, J.; Studeny, Z. Influence of Alloying Elements on Gas Nitriding Process of High-Stressed Machine Parts of Weapons. *Met. Mater.* **2018**, *56*, 97–103. [CrossRef]
27. Pokorny, Z.; Hruby, V.; Studeny, Z. Effect of nitrogen on surface morphology of layers. *Met. Mater.* **2016**, *54*, 119–124. [CrossRef]
28. Pokorny, Z.; Dobrocky, D.; Surlakova, M. Porosity in surface layers after chemical-heat treatment. In Proceedings of the Metal 2017, 25th Anniversary International Conference on Metallurgy and Materials, Brno, Czech Republic, 24–26 May 2017; TANGER Ltd.: Ostrava, Czech Republic, 2017; pp. 1147–1152.

materials

Article

Cyclic Fatigue of Dental NiTi Instruments after Plasma Nitriding

Michal Bumbalek [1], Zdenek Joska [2,*], Zdenek Pokorny [2], Josef Sedlak [3], Jozef Majerik [4], Vlastimil Neumann [2] and Karel Klima [1]

1 First Faculty of Medicine, Charles University Prague, 12108 Prague, Czech Republic; michalbumbalek@gmail.com (M.B.); karel.klima@lf1.cuni.cz (K.K.)
2 Faculty of Military Technology, University of Defence, 66210 Brno, Czech Republic; zdenek.pokorny@unob.cz (Z.P.); vlastimil.neumann@unob.cz (V.N.)
3 Faculty of Mechanical Engineering, Brno University of Technology, 61669 Brno, Czech Republic; sedlak@fme.vutbr.cz
4 Faculty of Special Technology, Alexander Dubcek University of Trencin, 91101 Trencin, Slovakia; jozef.majerik@tnuni.sk
* Correspondence: zdenek.joska@unob.cz; Tel.: +420-973-442-544

Abstract: This study investigated the possibility of nitride NiTi instruments using low-temperature plasma nitriding technology in a standard industrial device. Changes in the properties and fatigue life of used NiTi instruments before and after low-temperature nitriding application were investigated and compared. Nontreated and two series of plasma-nitrided NiTi instruments, designed by Mtwo company with tip sizes of 10/.04 taper, 15/.05 taper, and 20/.06 taper, were experimentally tested in this study. All these instruments were used and discarded from clinical use. The instruments were tested in an artificial canal made of stainless steel with an inner diameter of 1.5 mm, a 60° angle of curvature, and a radius of curvature of 3 mm. A low-temperature plasma nitriding process was used for the surface treatment of dental files using two different processes: 550 °C for 20 h, and 470 °C for 4 h. The results proved that it is possible to nitride dental instruments made of NiTi with a low-temperature plasma nitriding process. Promising results were achieved in trial testing by NiTi instruments nitrided at a higher temperature. Plasma-nitrided files were found to have, in some cases, significantly higher values than nontreated files in terms of fatigue life. The results showed that the nitriding process offers promising possibilities for suitably modified surface properties and quality of surface layer of NiTi instruments. Within the limitations of the present study, the cyclic fatigue life of plasma-nitrided NiTi dental files can be increased using this surface technology.

Keywords: nickel titanium; endodontic; plasma nitriding; cyclic fatigue; fracture

Citation: Bumbalek, M.; Joska, Z.; Pokorny, Z.; Sedlak, J.; Majerik, J.; Neumann, V.; Klima, K. Cyclic Fatigue of Dental NiTi Instruments after Plasma Nitriding. *Materials* **2021**, *14*, 2155. https://doi.org/10.3390/ma14092155

Academic Editor: Marcin Wachowski

Received: 21 March 2021
Accepted: 20 April 2021
Published: 23 April 2021

1. Introduction

Nickel titanium (NiTi) alloys are widely used in biomedical devices due to their unique mechanical properties, such as superelasticity, shape memory, and good biocompatibility. However, the release of nickel ions from the surface of nitinol is a problem. In addition to implantology, endodontics is also a branch of medicine where these NiTi alloys are very often used [1,2]. The development of endodontic nickel titanium instruments improves the treatment possibilities for root canals of teeth. NiTi instruments adapt better to the shape of the root canal, making treatment faster and with better repeatability than stainless steel instrument treatment [2–4]. Although NiTi alloy has excellent mechanical properties, clinical practice has shown that the risk of breakage with these instruments is greater than with traditional stainless steel tools. One of the most common causes of tool damage is cyclic fatigue [4–8]. It has been found that the fatigue life is affected by many factors, the most important of which are the shape of the tool cross-section, type and quality of tool production, size and length of the tool, and tool movement in the root canal [9–17]. Mechanical surface treatment is among the most frequently used methods of surface treatment of NiTi alloys and includes the processes of grinding, sandblasting, and polishing [18,19]. Chemical heat treatment of NiTi alloys includes nitriding, ion implantation, and oxidation processes. These surface treatments offer very promising results, as they form a protective abrasion-resistant layer of TiN in the case of nitriding, or TiO_2 in the case of oxidation, on the surface. Surface nitriding of NiTi alloys has been presented in many studies [20–27]. Ion implantation, gas nitriding, and powder nitriding are most often used to saturate the surface with nitrogen. These processes use a nitriding atmosphere of pure nitrogen [28–30] and a mixture of 90% nitrogen with 10% hydrogen [30]. All these processes take place at high temperatures in the range of from 800 °C to 1100 °C. The thickness of the layer varies depending on the nitriding time up to a thickness of 2 µm. Other types of nitriding include micropulse plasma nitriding. This type of process uses low temperatures in the nitriding process in the range of 300 °C to 580 °C. In this process, a mixture of nitrogen and hydrogen (N_2 and H_2) is most often used, where the molecules of these gases are cleaved and ionized in an electric field. Positive ions are accelerated towards the cathode, which is the surface of the nitrided components. Due to the sharp increase in the kinetic energy of the ions, the surface of the component is bombarded. Events occurring during the interaction between the plasma and the surface are shown in Figure 1.

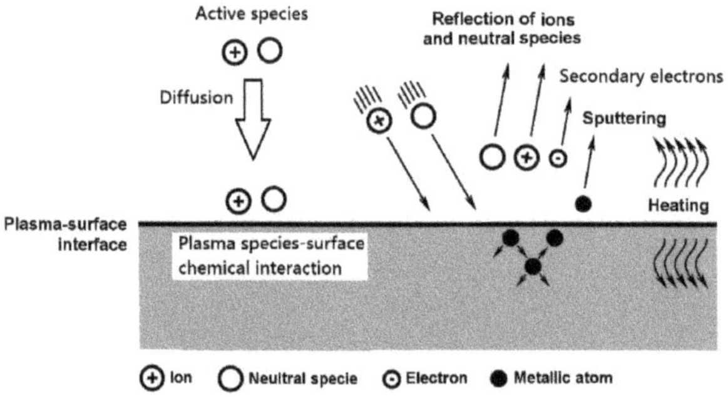

Figure 1. Illustration of events occurring during interaction between plasma and surface [31].

Upon impact, heating converts kinetic energy into heat; the components are heated; and, at the same time, atoms of other elements, especially carbon, oxygen, and nitrogen, are ejected from the surface. By applying ionic sputtering before nitriding, it is possible to remove the surface layer of oxides [25], reduce the nickel concentration in the top layer of the NiTi alloy [32], and, thus, increase the formation of diffuse layers of titanium nitride and reduce the temperature of the nitriding process to about 500 °C.

The aims of this study were to verify whether it is possible to form nitrided layers with endodontic instruments, which have very small dimensions and a complicated shape compared to standard nitrided parts, and to verify the effect of formed layers on the cyclic life of endodontic instruments.

2. Materials and Methods

The experiments were based on the testing of real instruments used in dental treatment. NiTi instruments, designed by Mtwo (VDW, Munich, Germany) with tip sizes of 10/.04 taper, 15/.05 taper, and 20/.06 taper, were experimentally tested. These three types of instruments were chosen because they are most commonly used in clinical practice for the treatment of dental root canals. For these experiments, discarded tools from clinical practice were used.

2.1. Surface Treatment

Prior to the plasma nitriding process, NiTi instruments were collected and randomly divided into three groups: nontreatment (benchmark), plasma nitrided at 470 °C for 4 h, and plasma nitrided at 550 °C for 20 h. Before treatment, the instruments were submitted to ultrasonic cleaning in two stages of ten minutes each (first in acetone and second in isopropyl alcohol). For this experiment, industrial plasma nitriding equipment Rubig PN 60/60 (Rubig, Wels, Austria) was used. For plasma nitriding, two processes that are used as standards in nitriding steel parts were chosen. Since there is no information in the literature about the plasma nitriding of real endodontic instruments, two "boundary processes" that can be performed by the nitriding device were chosen. Process No. 1, consisting of a low nitriding temperature of 470 °C and a short nitriding time of 4 h, and process No. 2, consisting of a high nitriding temperature of 550 °C and a long nitriding time of 20 h, were selected. The working atmosphere of process was 100% N_2. The concentration of N_2 in the working atmosphere was necessary due to the diffusion process from the outer environment to NiTi materials.

Diffusion is dependent on temperature, pressure, duration, and suitable alloying elements in the nontreated material. These materials are very difficult to nitride, especially due to the low affinity of Ni for nitrogen [32]. Before the nitriding process, the sputtering process was included. The parameters of plasma sputtering and plasma nitriding are given in Tables 1 and 2. The aims of the experiment were to compare the nitrided layers formed during these "boundary processes", with the lowest possible temperature used for nitriding in this device used over a short time and the highest temperature used for nitriding used over a long time, and to verify whether they have a different effect on the fatigue life of endodontic instruments.

Table 1. Parameters of plasma sputtering.

Plasma Sputtering					
Temperature (°C)	**Time (h)**	**Gas Flow H_2 ($l \cdot min^{-1}$)**	**Bias (V)**	**Pressure (Pa)**	**Pulse Length (μs)**
450	0.5	8	800	80	100

Table 2. Parameters of plasma nitriding processes.

	Plasma Nitriding					
	Temperature (°C)	**Time (h)**	**Gas Flow N_2 ($l \cdot min^{-1}$)**	**Bias (V)**	**Pressure (Pa)**	**Pulse Length (μs)**
Process No. 1	470	4	8	800	80	100
Process No. 2	550	20	8	800	80	100

2.2. Chemical Composition

The chemical composition of the NiTi instruments was verified using the EDS method on a TESCAN MIRA 4 electron microscope. Line scan measurements of the chemical composition of the plasma-nitrided surface and the untreated surface were used. It is often stated in the literature [21,22] that the nickel content in the NiTi alloy is in the range of 48–52 wt %. During use, the chemical composition of the elements on the surface changes [23,24]; the nickel content decreases; and the content of titanium and other elements, such as nitrogen or oxygen, increases.

2.3. Sample Preparation for Evaluation of Nitrided Layer

The presence of a nitride layer in NiTi material was evaluated with the scanning electron microscopy method by using a TESCAN MIRA 4. The evaluation of the thickness of the nitride layer was performed on a cross-section. The samples of instruments were pressed into thermoplastic powder, subsequently ground, and, finally, polished. During the grinding process, the samples were checked step by step using optical microscopy. Finally, the samples were wet ground with sandpaper and polished using velvet with a diamond paste with grains 1 μm in size. The samples' surface, thus prepared, was etched in an acid bath (HNO_3: HF: CH_3COOH 1:2:7). For the observation of the microstructure, the Olympus DSX 500i (Olympus, Tokyo, Japan) optical microscope with a magnification of 1000× was utilized. The evaluation of the fracture was performed on an SEM TESCAN MIRA 4 (Tescan, Brno, Czech Republic) without a special sample preparation process.

2.4. Cyclic Fatigue Test

Cyclic fatigue testing in artificial canals was performed with an electric motor (Endo a class reciprocating LED, Medin, Czech Republic). The simulated canals were machined from a stainless steel block. The simulated canal was set as R 3 mm with an angle of 60° (Figure 2). The rotation speed was 190 rpm without torque control. The fracture time was recorded for each file in seconds and repeated ten times (n = 10). The corresponding fatigue life in the number of cycles was then accurately calculated.

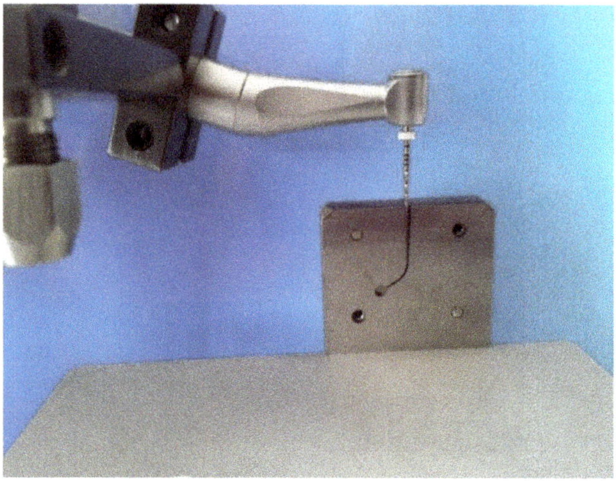

Figure 2. Testing stand of cyclic fatigue of NiTi instruments in artificial canal.

3. Results and Discussion

3.1. Chemical Composition

The chemical composition of dental NiTi instruments was verified using the EDS method due to a very small measurable area on the physical sample, as shown in Figure 3. The use of other typical spectroscopy testing methods was not possible.

The concentrations of Ni, Ti, and N elements were obtained via the EDS method using SEM TESCAN MIRA 4, and the results of the measurements are displayed in Table 3.

Table 3. Chemical composition of untreated and plasma nitrided surfaces of NiTi instrument verified with EDS EDAX (wt. %).

	Ni	Ti	N
		EDS	
Untreated surface	46	53	0
Surface after Process No. 1	42	49	9
Surface after Process No. 2	32	54	14

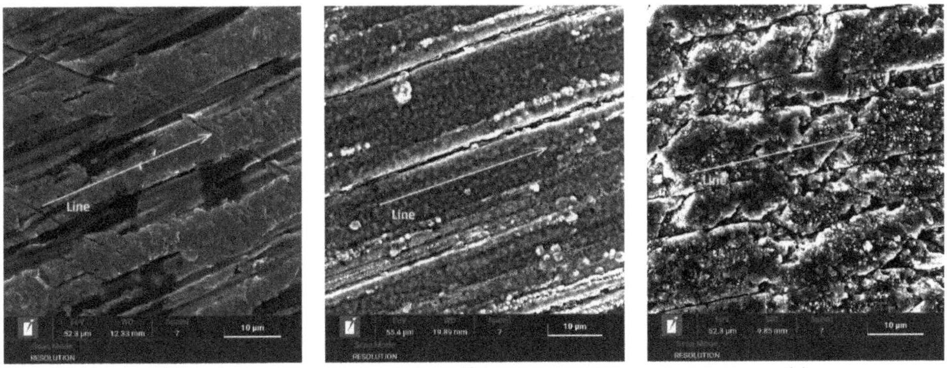

(a) (b) (c)

Figure 3. EDS line scan of untreated surface (**a**), plasma-nitrided surface after plasma nitriding at 470 °C and 4 h (**b**), plasma-nitrided surface after plasma nitriding at 550 °C and 20 h (**c**).

The content of elements measured by the analysis is in good agreement with the values measured in other studies [22–24]. The nickel content decreased in both plasma nitriding processes compared to the measurement performed on the untreated surface. The nitrogen content was higher in the plasma nitriding Process No. 2 and reached a value of 14 wt. %. It is well-known that alloys with a high concentration of Ni, Ti, Co, and Mn are very difficult to nitride [33]. The concentrations of Ni and Ti elements in this alloy are at a very high level (Table 3), and, therefore, the process of nitriding must be optimized in the future for supporting the diffusion process in this specific NiTi material during the process.

3.2. Assessment of Nitrided Layer

The surface documentation was provided using optical microscopy on an Olympus DSX 100 device. After plasma nitriding in both processes, the surface of the instruments changed color from metallic silver to dark bronze in Process No. 1 and to dark brown in Process No. 2, which indicates the probable presence of titanium nitrides, as seen in Figure 4a. Figure 4b shows an example of tool tips after the bending test of NiTi instruments. The untreated instrument showed a memory effect typical for these NiTi instruments, while, after both plasma nitriding processes, this memory effect was lost.

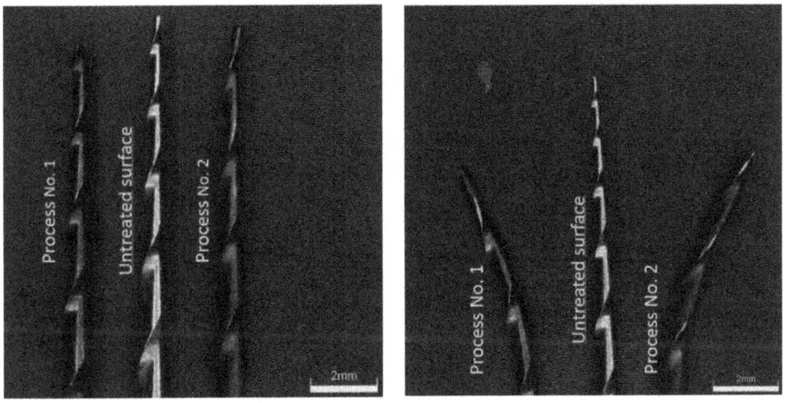

(a) (b)

Figure 4. Macroscopic image of color change in the instrument with a tip size of 20/.06 taper prior and after plasma nitriding at 550 °C for 20 h (**a**) and rigid tip in the instrument with a tip size of 15/.05 taper after plasma nitriding at 550 °C for 20 h (**b**).

Figures 5 and 6 show the structure of the nitrided layer after the processes, documented using SEM. The nitrided layers after both processes had a variable thickness: in the case of Process No. 1, the thickness of the nitrided layer reached 2–2.7 μm, and, in the case of Process No. 2, the thickness of the nitrided layer reached 3–3.8 μm.

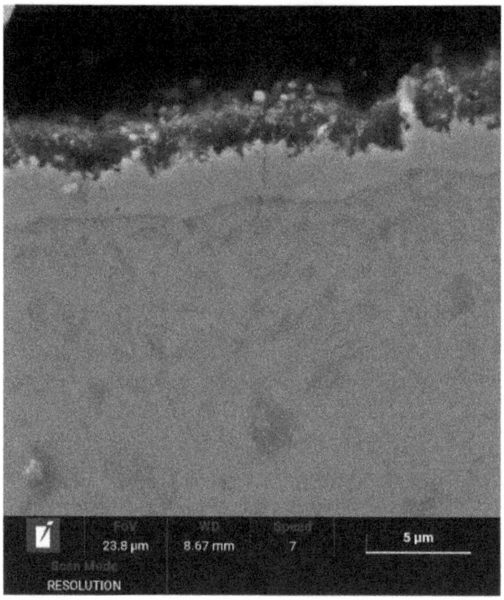

Figure 5. Nitrided layer on NiTi instrument with a tip size of 20/.06 taper after plasma nitriding at 470 °C and 4 h.

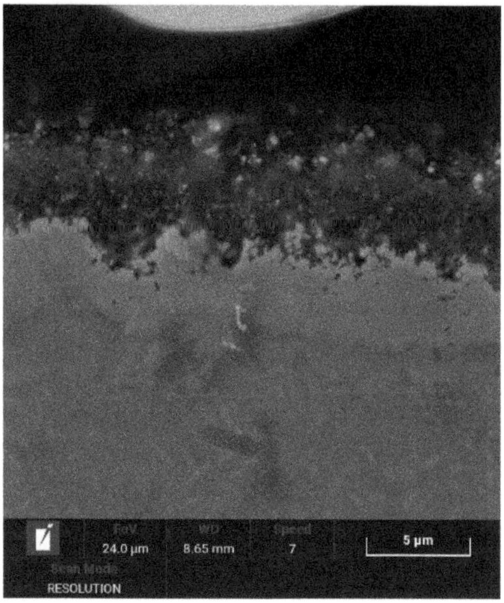

Figure 6. Nitrided layer on NiTi instrument with a tip size of 15/.05 taper after plasma nitriding at 550 °C for 20 h.

Figure 7a,b show the difference between the microstructure of NiTi after plasma nitriding. Figure 7a shows a fine martensitic structure, while Figure 7b shows a markedly coarse martensitic structure.

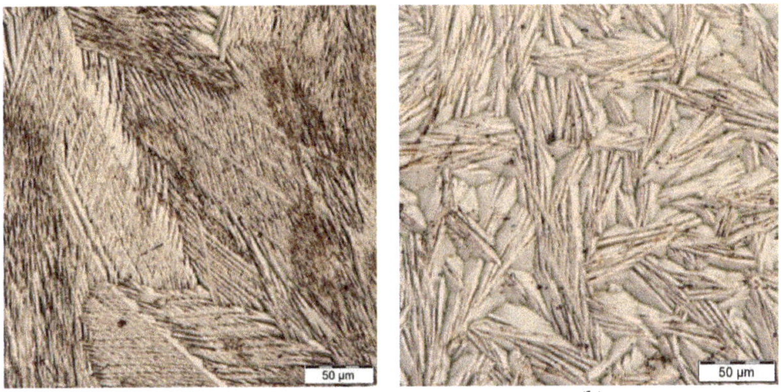

(**a**) (**b**)

Figure 7. Microscopic image of different martensitic structures of instruments after plasma nitriding process 470 °C for 4 h (**a**) and plasma nitriding process 550 °C for 20 h. (**b**).

3.3. Cyclic Fatigue Test

A simple cyclic fatigue test was performed to determine if the formed nitride layer affected the residual life of the instrument. In the cyclic fatigue test, there were significant differences in the cyclic fatigue of the untreated and nitrided instruments. Figure 8 Represents the comparison of individual tools after cyclic fatigue testing. The blue bars show new [34], unused instruments, which were also tested for comparison as the establishment of an initial state. For instruments with a tip size of 20/.06 taper and a tip size of 15/.05 taper, values were similar to the 420 cycles to fracture. For the tip size 10/.04 taper instruments, the number of cycles was 570 to breakage. For the instruments used (orange bar), the residual life in the raw state was measured. For instruments of all sizes, the results showed that instruments discarded from clinical use have half the service life of new instruments. For plasma-nitrided instruments after Process No. 1, the cyclic tool life of all three tool types was reduced. The most significant reduction occurred with the instrument with a tip size of 15/.05 taper. For plasma-nitrided tools in Process No. 2, the instrument with a tip size of 15/.05 taper had a small reduction in cyclic fatigue life compared to the instrument used. After Process No. 2, the cyclic fatigue life of the instrument with a tip size of 20/.06 taper increased, with the values being even higher than the new instrument. For the instrument with a tip size of 10/.04 taper, there was a marked increase in the residual cyclic life after Process No. 2 where the values reached 650 cycles compared to the 480 cycles reached by new instruments.

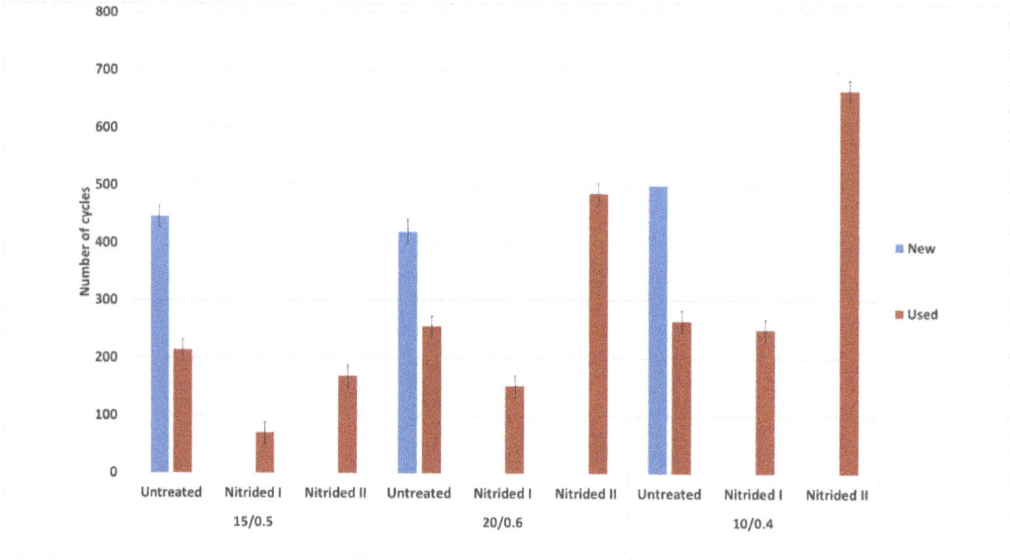

Figure 8. Results of cyclic fatigue test.

3.4. Fractographic Analysis

In both plasma nitriding processes, increased rigidity of the instruments was observed, where, during the experiment, it was evident from Figure 4b that the tool tip did not return to its original position. This can only be associated with the annealing of the NiTi base material during plasma nitriding, where the martensitic needles of the NiTi instrument coarsened. Fractographic analysis was performed with an electron microscope, TESCAN MIRA 4. In Figures 9 and 10, the difference between the fractures of individual plasma-nitrided instruments is visible. The fractured character of the plasma-nitrided instruments can be seen to be changed and formed by a mix of ductile and brittle fractures. In plasma-nitrided tools, creek-shaped reliefs of the crack propagation process are visible. Furthermore, it can be seen that the microstructure of the fracture surface is different and is in good agreement with Figure 7a,b; in Process No. 1, the structure of the fracture surface is finer than that in Process No. 2. The coarsening of the instrument grain occurred because the nitriding process took place at a higher temperature of 550 °C and the cooling process in the nitriding device was slow. In contrast, nitriding at 470 °C did not cause such significant grain coarsening.

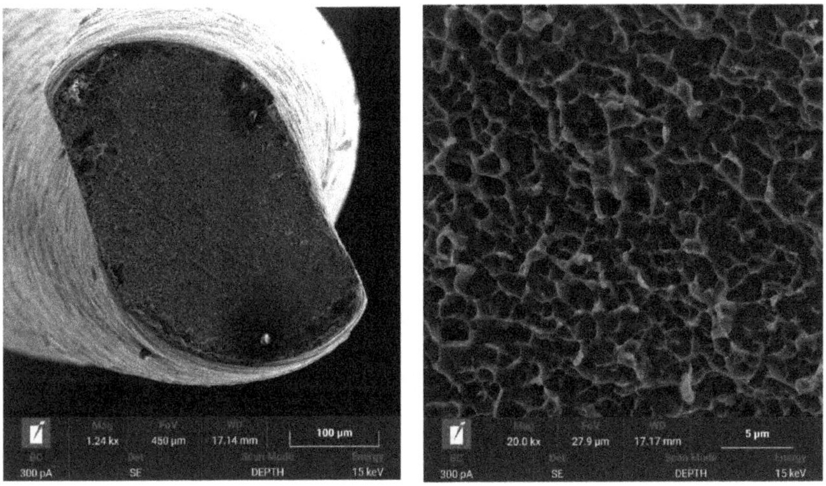

(**a**) (**b**)

Figure 9. Macroscopic image of instrument with a tip size of 15/0.5 taper after plasma nitriding at 470 °C for 4 h (**a**) and detailed fracture surface (**b**).

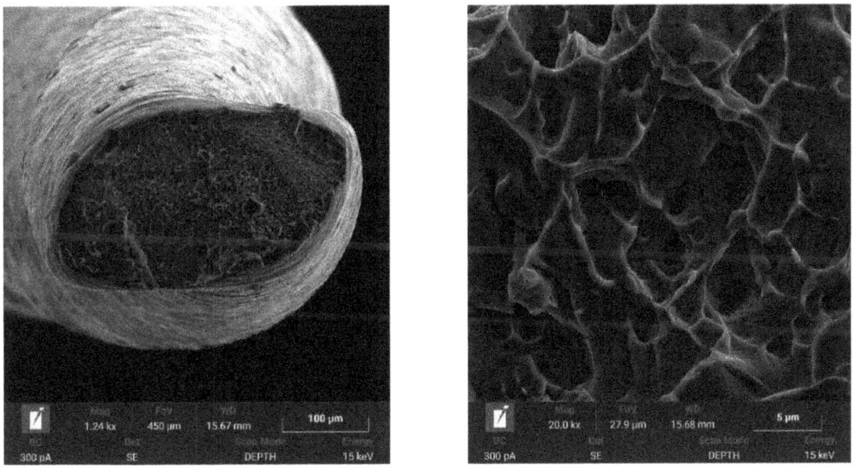

(**a**) (**b**)

Figure 10. Macroscopic image of instrument with a tip size of 15/.05 taper after plasma nitriding at 550 °C for 20 h (**a**) and detailed fracture surface (**b**).

4. Conclusions

The nitrided layer in Process No. 1 reached a thickness of about 2–2.7 µm. In Process No. 2, there was an increase in nitrided layer thickness of 3–3.8 µm and in the size of martensitic needles in the base material. This was reflected in the nature of the tool fracture, where the shape and profile of the fracture surface was mixed with a large proportion of ductile fracture. In Process No. 1 there were finer martensitic needles in the base material, and the fracture surface did not show as high a proportion of ductile fracture. In both cases, with a more detailed observation of the fracture surfaces in all cases, the crack propagated along the grain boundaries. The results of cyclic fatigue were not unambiguous for nitrided instruments. For instruments with a tip size of 20/.06 taper and the instrument with a tip size of 10/.04 taper for Process No. 1, the number of cycles to fracture was reduced. Nitriding Process No. 2 had significantly better results with both tools, as the fracture speed was almost double that of the tools used, and even better than the new tools. For the instrument with a tip size of 15/.05 taper, the results were not unambiguous; for Process No. 1, there was a significant reduction in the number of cycles to fracture, and in Process No. 2, there was a partial reduction in the number of cycles to fracture. Based on these simple results, there is room for further investigation and optimization of plasma nitriding processes for endodontic instruments. By optimizing plasma sputtering, the composition of the nitriding atmosphere, voltage, and length of the micropulse discharge, it will be possible to create a suitable nitrided layer on endodontic instruments in the future.

Within the limitations of this study, nitrided endodontic instruments showed more promising physical and mechanical properties than conventional endodontic instruments. Future research should focus on new raw materials and thermomechanical processing procedures that can be used to further optimize the alloy microstructure in order to guarantee the reliability and safety of NiTi rotary instruments in clinical practice.

Author Contributions: Conceptualization, Z.J., J.M. and Z.P.; methodology, Z.J., Z.P. and M.B.; formal analysis, Z.J., V.N.; investigation, Z.J., K.K., J.S.; resources, Z.J., Z.P., M.B. and J.S.; writing—original draft preparation, Z.J. and M.B.; writing—review and editing, Z.P., M.B. and J.M.; supervision, Z.P., V.N. and M.B. All authors have read and agreed to the published version of the manuscript.

Funding: The work presented in this paper has been supported by the specific research project 2020 "SV20-216"at the Department of Mechanical Engineering, University of Defence in Brno and the Project for the Development of the Organization "DZRO Military autonomous and robotic systems" and the Slovak Research and Development Agency under contract No. APVV-15-0710.

Institutional Review Board Statement: Not applicable.

Informed Consent Statement: Not applicable.

Data Availability Statement: Data are contained within the article.

Conflicts of Interest: The authors declare no conflict of interest.

References

1. Torrisi, L. The NiTi superelastic alloy application to the dentistry field. *BioMed. Mater. Eng.* **1999**, *9*, 39–47. [PubMed]
2. Thompson, S.A. An overview of nickel-titanium alloys used in dentistry. *Int. Endod. J.* **2000**, *33*, 297–310. [CrossRef]
3. Cheung, G.S.P.; Zhang, E.W.; Zheng, Y.F. A numerical method for predicting the bending fatigue life of NiTi and stain-less steel root canal instruments. *Int. Endod. J.* **2011**, *44*, 357–361. [CrossRef] [PubMed]
4. Plotino, G.; Grande, N.M.; Cordaro, M.; Testarelli, L.; Gambarini, G. A Review of Cyclic Fatigue Testing of Nickel-Titanium Rotary Instruments. *J. Endod.* **2009**, *35*, 1469–1476. [CrossRef] [PubMed]
5. Alrahabi, M. Comparative study of root-canal shaping with stainless steel and rotary NiTi files performed by preclinical dental students. *Technol. Health Care* **2015**, *23*, 257–265. [CrossRef]
6. Parashos, P.; Gordon, I.; Messer, H.H. Factors Influencing Defects of Rotary Nickel-Titanium Endodontic Instruments after Clinical Use. *J. Endod.* **2004**, *30*, 722–725. [CrossRef]
7. Schirrmeister, J.F.; Strohl, C.; Altenburger, M.J.; Wrbas, K.T.; Hellwig, E. Shaping ability and safety of five different rotary nick-el-titanium instruments compared with stainless steel hand instrumentation in simulated curved root canals. *Oral Surg. Oral Med. Oral Pathol. Oral Radio Endod.* **2006**, *101*, 807–813. [CrossRef] [PubMed]
8. Spili, P.; Parashos, P.; Messer, H.H. The Impact of Instrument Fracture on Outcome of Endodontic Treatment. *J. Endod.* **2005**, *31*, 845–850. [CrossRef]
9. Li, U.M.; Lee, B.S.; Shih, C.T.; Lan, W.H.; Lin, C.P. Cyclic fatigue of endodontic nickel titanium rotary instruments: Static and dy-namic tests. *J. Endod.* **2002**, *28*, 448–451. [CrossRef]
10. Tripi, T.R.; Bonaccorso, A.; Condorelli, G.G. Cyclic fatigue of different nickel-titanium endodontic rotary instruments. *Oral Surg. Oral Med. Oral Pathol. Oral Radiol. Endod.* **2006**, *102*, e106–e114. [CrossRef]
11. Berutti, E.; Chiandussi, G.; Gaviglio, I.; Ibba, A. Comparative Analysis of Torsional and Bending Stresses in Two Mathematical Models of Nickel-Titanium Rotary Instruments: ProTaper versus ProFile. *J. Endod.* **2003**, *29*, 15–19. [CrossRef] [PubMed]
12. Turpin, Y.L.; Chagneau, F.; Vulcain, J.M. Impact of two theoretical cross-sections on torsional and bending stresses of nickelti-tanium root canal instrument models. *J. Endod.* **2000**, *26*, 414–417. [CrossRef] [PubMed]

13. Hornbogen, E. Some effects of martensitic transformation on fatigue resistance. *Fatigue Fract. Eng. Mater. Struct.* **2002**, *25*, 785–790. [CrossRef]
14. Gavini, G.; Dos Santos, M.; Caldeira, C.L.; Machado, M.E.D.L.; Freire, L.G.; Iglecias, E.F.; Peters, O.A.; Candeiro, G.T.D.M. Nickel–titanium instruments in endodontics: A concise review of the state of the art. *Braz. Oral Res.* **2018**, *32*, 67. [CrossRef] [PubMed]
15. Loska, S.; Basiaga, M.; Pochrząst, M.; Łukomska-Szymańska, M.; Walke, W.; Tyrlik-Held, J. Comparative characteristics of endodon-tic drills. *Acta Bioeng. Biomech.* **2015**, *17*, 75–83.
16. Hamdy, T.M.; Galal, M.; Ismail, A.G.; Abdelraouf, R.M. Evaluation of Flexibility, Microstructure and Elemental Analysis of Some Contemporary Nickel-Titanium Rotary Instruments. *Open Access Maced. J. Med. Sci.* **2019**, *7*, 3647–3654. [CrossRef] [PubMed]
17. Bastos, M.M.B.; Hanan, A.R.A.; Marques, A.A.F.; Garcia, L.D.F.R.; Júnior, E.C.S. Topographic and Chemical Analysis of Reciprocating and Rotary Instruments Surface after Continuous Use. *Braz. Dent. J.* **2017**, *28*, 461–466. [CrossRef]
18. Anderson, M.E.; Price, J.W.; Parashos, P. Fracture Resistance of Electropolished Rotary Nickel–Titanium Endodontic Instruments. *J. Endod.* **2007**, *33*, 1212–1216. [CrossRef] [PubMed]
19. Sharma, S.; Kumar Tewari, R.; Kharade, P.; Kharade, P. Comparative Evaluation of the Effect of Manufacturing Process on Dis-tortion of Rotary ProFile and Twisted File: An in Vitro SEM Study. *J. Dent. Res. Dent. Clin. Dent. Prospect.* **2015**, *9*, 216–220. [CrossRef]
20. Gavini, G.; Pessoa, O.F.; Barletta, F.B.; Vasconcellos, M.A.; Caldeira, C.L. Cyclic fatigue resistance of rotary nickel titanium instru-ments submitted to nitrogen ion implantation. *J. Endod.* **2010**, *36*, 1183–1186. [CrossRef] [PubMed]
21. Zhao, N.; Man, H.; Cui, Z.; Yang, X. Structure and wear properties of laser gas nitrided NiTi surface. *Surf. Coat. Technol.* **2006**, *200*, 4879–4884. [CrossRef]
22. Dos Santos, M.; Gavini, G.; Siqueira, E.L.; Da Costa, C. Effect of Nitrogen Ion Implantation on the Flexibility of Rotary Nickel-Titanium Instruments. *J. Endod.* **2012**, *38*, 673–675. [CrossRef]
23. Shevchenko, N.; Pham, M.-T.; Maitz, M. Studies of surface modified NiTi alloy. *Appl. Surf. Sci.* **2004**, *235*, 126–131. [CrossRef]
24. Wolle, C.F.; Vasconcellos, M.A.; Hinrichs, R.; Becker, A.N.; Barletta, F.B. The effect of argon and nitrogen ion implantation on nickeletitanium rotary instruments. *J. Endod.* **2009**, *35*, 1558–1562. [CrossRef]
25. Al Jabbari, Y.S.; Fehrman, J.; Barnes, A.; Zapf, A.; Zinelis, S.; Berzins, D. Titanium nitride and nitrogen ion implanted coated dental materials. *Coatings* **2012**, *2*, 160–178. [CrossRef]
26. Firstov, G.; Vitchev, R.; Kumar, H.; Blanpain, B.; Van Humbeeck, J. Surface oxidation of NiTi shape memory alloy. *Biomaterials* **2002**, *23*, 4863–4871. [CrossRef]
27. Aun, D.P.; Peixoto, I.F.D.C.; Houmard, M.; Buono, V.T.L. Enhancement of NiTi superelastic endodontic instruments by TiO2 coat-ing. *Mater. Sci. Eng. C Mater. Biol. Appl.* **2016**, *68*, 675–680. [CrossRef] [PubMed]
28. Gil, F.J.; Solano, E.; Campos, A.; Boccio, F.; Sáez, I.; Alfonso, M.V.; Planell, J.A. Improvement of the friction behaviour of NiTi ortho-dontic archwires by nitrogen diffusion. *Biomed. Mater Eng.* **1998**, *8*, 335–342. [PubMed]
29. Vojtech, D.; Fojt, J.; Joska, L.; Novak, P. Surface treatment of NiTi shape memory alloy and its influence on corrosion behavior. *Surf. Coat. Technol.* **2010**, *204*, 3895–3901. [CrossRef]
30. Lutz, J.; Lindner, J.; Mändl, S. Marker experiments to determine diffusing species and diffusion path in medical Nitinol alloys. *Appl. Surf. Sci.* **2008**, *255*, 1107–1109. [CrossRef]
31. Rodrigo, P.C.; Marcio, M.; Silvio, F.B. Low-Temperature Thermochemical Treatments of Stainless Steels—An Introduction. In *Plasma Science and Technology—Progress in Physical States and Chemical Reactions*; Mieno, T., Ed.; IntechOpen: London, UK, 20 April 2016. [CrossRef]
32. Lelątko, J.; Goryczka, T.; Wierzchoń, T.; Ossowski, M.; Łosiewicz, B.; Rówiński, E.; Morawiec, H. Structure of Low Temperature Nitrided/Oxidized Layer Formed on NiTi Shape Memory Alloy. *Solid State Phenom.* **2010**, *163*, 127–130. [CrossRef]
33. Prochazka, J.; Pokorny, Z.; Dobrocky, D. Service Behavior of Nitride Layers of Steels for Military Applications. *Coatings* **2020**, *10*, 975. [CrossRef]
34. Fišerová, E.; Chvosteková, M.; Bělašková, S.; Bumbálek, M.; Joska, Z. Survival Analysis of Factors Influencing Cyclic Fatigue of Nickel-Titanium Endodontic Instruments. *Adv. Mater. Sci. Eng.* **2015**, *2015*, 189703. [CrossRef]

materials

MDPI

Article

Physical and Mechanical Properties of Polypropylene Fibre-Reinforced Cement–Glass Composite

Marcin Małek [1], Waldemar Łasica [1], Marta Kadela [2,*], Janusz Kluczyński [3] and Daniel Dudek [2]

[1] Faculty of Civil Engineering and Geodesy, Military University of Technology in Warsaw, ul. Gen. Sylwestra Kaliskiego 2, 01-476 Warsaw, Poland; marcin.malek@wat.edu.pl (M.M.); waldemar.lasica@wat.edu.pl (W.Ł.)
[2] Building Research Institute (ITB), ul. Filtrowa 1, 00-611 Warsaw, Poland; d.dudek@itb.pl
[3] Faculty of Mechanical Engineering, Military University of Technology, ul. Gen. Sylwestra Kaliskiego 2, 00-908 Warsaw, Poland; janusz.kluczynski@wat.edu.pl
* Correspondence: m.kadela@itb.pl; Tel.: +48-603-60-12-48

Abstract: In accordance with the principles of sustainable development, environmentally friendly, low-emission, and energy-intensive materials and technologies, as well as waste management, should be used. Concrete production is responsible for significant energy consumption and CO_2 production; therefore, it is necessary to look for new solutions in which components are replaced by other materials, preferably recycled. A positive way is to use glass waste. In order to determine the effect of a significant glass cullet content on the properties of concrete, glass powder was used as a filler and 100% glass aggregate. The cement–glass composite has low tensile strength and brittle failure. In order to improve tensile strength, the effects of adding polypropylene fibres on the mechanical properties of the composite were investigated. With the addition of 300, 600, 900, 1200, and 1500 g/m^3 of fibres, which corresponds to 0.0625%, 0.1250%, 0.1875%, 0.2500%, and 0.3125% of cement mass, respectively, flexural strength increased compared with the base sample by 4.1%, 8.2%, 14.3%, 20.4%, and 26.5%, respectively, while the increase in splitting strength was 35%, 45%, 115%, 135%, and 185%, respectively. Moreover, with the addition of fibres, a decrease in slump by 25.9%, 39.7%, 48.3%, 56.9%, and 65.5%, respectively, compared with the reference specimen was determined.

Keywords: by-product waste; packaging waste; glass cullet; macro-polymeric fibre; recycling; eco-efficient concrete; slump cone; compressive strength; flexural strength; splitting strength

Citation: Małek, M.; Łasica, W.; Kadela, M.; Kluczyński, J.; Dudek, D. Physical and Mechanical Properties of Polypropylene Fibre-Reinforced Cement–Glass Composite. *Materials* **2021**, *14*, 637. https://doi.org/10.3390/ma14030637

Academic Editor: Hyeong-Ki Kim
Received: 30 November 2020
Accepted: 24 January 2021
Published: 30 January 2021

Highlights

- Recycled macro-polymer fibres were used to improve tensile strength of the cement–glass composite;
- Reduction of the workability of the cement–glass composite with the addition of polypropylene fibres was obtained;
- Slight effect of waste fibres on the compressive strength of the cement–glass composite was determined;
- With the addition of polypropylene fiber, the flexural strength of the composite increased;
- Significant increase in splitting strength for the fibre-reinforced cement–glass composite was demonstrated.

1. Introduction

The main problem of recent times is environmental pollution. This is related to, among others, an increase in the production and consumption of polymer materials by an average of 9% per year and an assumption that the upward trends will be continued (it is estimated that, in the coming years, the increase will be 5% a year [1]). According to Plastics Europe [2], world polymer production increased from 1.5 million tonnes in 1950 to 245 million tonnes in 2008. The European Union (EU) economy produces around 20% of the total world polymer production. The demand for raw materials in individual

European countries strongly depends on the size of the given economy and the degree of its development. For example, economically leading countries such as Germany, Italy, France, and the United Kingdom consume as many polymer materials as all other European countries put together [3]. According to Plastic Europe Polska [4], however, consumption has been constant for individual countries in recent years. Only the demand for individual types of polymer materials is subject to fluctuations [5]. An exception is, among others, Poland, where demand increases every year, regardless of the global economic situation [4]. According to Forbes [6], the global demand for certain plastic uses has increased owing to the threat posed by coronavirus. These include mainly polymer polypropylene, used in takeout food packaging, and polyethylene terephthalate (PET) in single-use plastic water bottles. This was the result of the shift from sit-in restaurants to take-out delivery and of the stockpiling of groceries and bottled water by consumers. For the same reason, the amount of glass waste increased, which is one of the most common materials in everyday life. The recycling rate of glass waste is quite low in many countries, compared with other solid wastes [7]. For example, in the United States, 11.38 million tonnes of waste glass were produced in 2017, but 26.6% was recycled and mainly used for the production of containers and packing, and 60.37% was landfilled [8]. In Hong Kong, 4063 and 7174 tonnes of glass waste was generated in 2018 and 2019 respectively, and the recovery rate was about 16.3% in 2018. The total amount of used glass containers that ended up in landfills in 2018 was 77,400 tonnes [9]. In Singapore, 72.8 million tonnes of glass were disposed in 2011, but only 29% was recycled [7]. In the United Kingdom, 1.85 million tonnes of waste glass are collected annually, and for container glass, the municipal recycling rate is 34% [10].

Concrete is a material that allows for the disposal of waste [11–20]. This is very important in the case of non-biodegradable or hardly decomposable materials such as polymers or glass waste. Polymer materials have been used as a replacement for natural aggregate in concrete [21–25], replacement of cement [26–28], additions (e.g., PET bottles [29–31], polyvinyl chloride (PVC) pipes [32], high density polyethylene (HDPE) [33], and thermosetting plastics [34]), expanded polystyrene foam (EPS) [35], polypropylene fibres [31,36,37], admixtures (e.g., polycarbonate and polyurethane foam [38–40]), or elements to concrete (e.g., concrete reinforcing bars [41] and plastic anchors [3,42,43]). The used polypropylene fibre by-products of recycling plastic packaging in concrete compared with plain concrete have been discussed in detail in a previous study [44]. Glass cullet may be used as a replacement for cement or aggregate [45–48], while the pozzolanic reactivity of glass powder with particle size below 100 μm is observed as an increase in compressive strength [49–51]. The impact of glass powder as a cement replacement on concrete or geopolymer properties was presented in [52–56]. For example, Federico and Chidiac [57] analysed the kinetic and performance properties of cementitious mixes with glass powder. Mirzahosseini and Riding [58] investigated the impact of curing temperature and glass type on the pozzolanic reaction and properties of concrete with glass powder.

Many scientists have tested concrete with glass aggregate as a replacement of coarse aggregate, fine aggregate, or cement in order to use waste glass in the concrete industry [59–61]. Yu et al. [62] reported that the glass cullet used as aggregate in concrete enhanced its mechanical properties. Limbachiya et al. [63] and Tittarelli et al. [64], however, obtained the same mechanical performances for concrete mixes with addition of glass sand up to 15%. It was found that the use of glass cullet as replacement for coarse aggregate is not satisfactory owing to the reduction of the bonding between the aggregate and the cement matrix, and a reduction of strength [65]. The effect of the size of glass particles on the properties of fresh mix and hardened samples was analysed by Ling and Poon [66] and Yousefi et al. [67]. However, the impact of fibres on the properties of a cement–glass composite has rarely been reported.

On the other hand, the production of building materials is responsible for significant energy consumption and CO_2 production [68–70], so it is necessary to look for new materials that can replace the currently used ones [71,72], preferably recycled [73,74]. The addition of recycled glass aggregate in concrete as a replacement of 5%, 10%, and 15% natural aggregate has been studied in previous research of the authors [75]. In this study, 100% of glass aggregate is used, which is a continuation and extension of research conducted by Małek at el. [74]. Moreover, as the cement–glass composite has low tensile strength and brittle failure, polypropylene fibres are additionally added to improve its tensile strength. In order to contribute to the use of recycled materials, polypropylene fibres were made from post-consumer waste (food packaging). This research aims to assess the influence of different fibre content on the mechanical properties of the cement–glass composite.

2. Materials

2.1. Products of Cementitious Mix

In this research, Portland cement, tap water, and polycarboxylate superplasticizer were used. The concept of designing a cement–glass composite is based on a single binder in the form of white cement. Because of the fact that the composition of the composite consisted of a 100% granulated glass cullet, the industrial white Portland cement CEM I 52.5R NA, pH = 13 was used. It is a special purpose cement with a very low content of alkaline compounds. The chemical composition of cement was investigated by PN EN 196-6:2011 [76] and is presented in Table 1. The physical properties and compressive strength of cement were determined according to PN-EN 196-6:2011 [76] and PN EN 196-1:2016-07 [77], respectively (see Table 2). The shape and texture of cement gain particle were investigated by scanning electron microscopy (SEM—Joel JSM 6600, Yvelines, France), as shown in Figure 1.

Table 1. Chemical composition of cement and glass cullet [78,79].

Compositions		SiO_2	Al_2O_3	Fe_2O_3	CaO	MgO	SO_3	Na_2O	K_2O	TiO_2	Cl
Unit (vol.%)	Cement	19.5	4.9	2.9	63.3	1.3	2.8	0.1	0.9	-	0.05
	Glass	70.0–74.0	0.5–2.0	0.0–0.1	7.0–11.0	3.0–5.0	-	13.0–15.0		0.0–0.1	-

Table 2. Physical properties of cement and glass cullet [79].

Properties	Specific Surface Area [m²/kg]	Specific Gravity [kg/m³]	Compressive Strength after Days [MPa]		
Materials			2 Days	7 Days	28 Days
Cement	400	3090–3190	40–48	53–65	66–76
Glass	100	2450	-	-	-

Figure 1. Scanning electron microscopy (SEM) image of used cement.

Recycled sodium glass granules were used as aggregate in the composition of the composite. Sieve analysis was performed using the "dry" method for three samples of the granulate using a laboratory shaker with a set of standard sieves with square meshes made of calibrated mesh. The percentage distribution of individual fractions of screened granules was determined. The crumb pile was designed from two granules of the fraction groups 0/0.9 (0/1) and 0.9/1.5 (1/2). As an additive in the composition of the composite (filler), glass powder from sodium glass with a particle size from 0 to100 µm and dry density of 1.0 kg/m^3 was used. The glass powder acts as a sealer for the crumb pile, the purpose of which is to ensure the continuity of the internal structure of the material. The graining curve was designed following the graining guidelines for sand concrete; the designed curve was related to the upper and lower limit curves. The designed curve ran in the area of good particle size distribution. The gradation curve of the used glass aggregate is presented in Figure 2. The crushed glass cullet revealed sharp edges, a rougher surface texture, and no cracks (Figure 3). The specific density and Mohs hardness scale of the glass aggregate was approximately 1.6 kg/m^3 and 6–7, respectively. The fineness modulus of the glass sand aggregate was equal to 2.56 MPa. The chemical composition and physical properties of glass cullet are given in Tables 1 and 2, respectively.

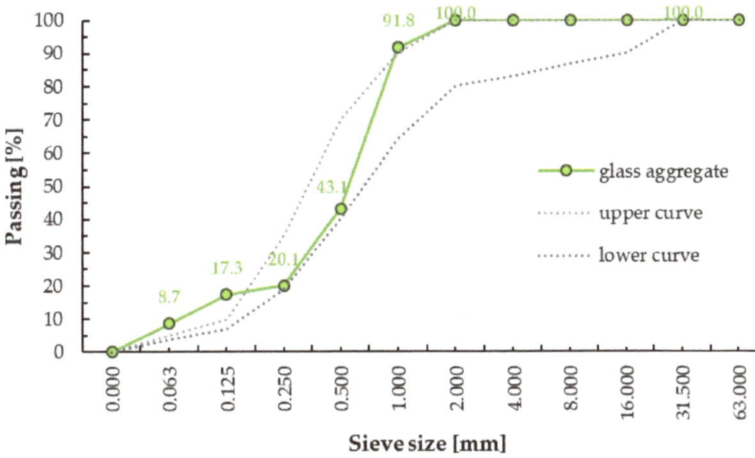

Figure 2. Gradation curve of glass aggregate.

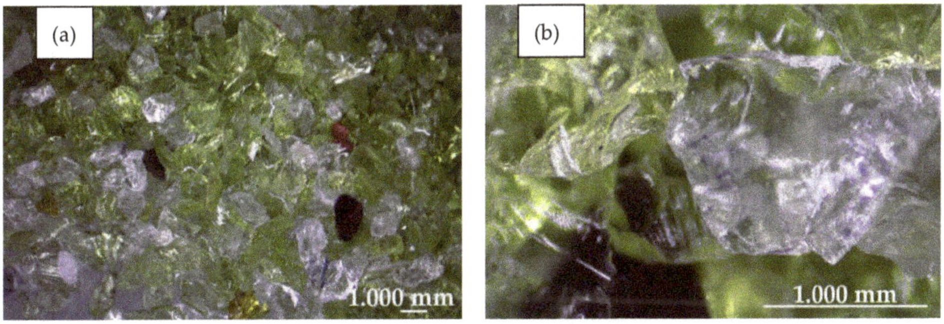

Figure 3. Light microscope images of used glass aggregate. (**a**) magnification of x5; (**b**) magnification of x45.

A superplasticizer was used in this research based on modified polycarboxylate ethers (melamine and silanes/siloxanes). The superplasticizer was added to reduce the amount of water (maintaining a water/cement ratio w/c at 0.26). The particle shape and texture of the admixture were investigated by scanning electron microscopy (Joel JSM 6600, Yvelines, France), as shown in Figure 4.

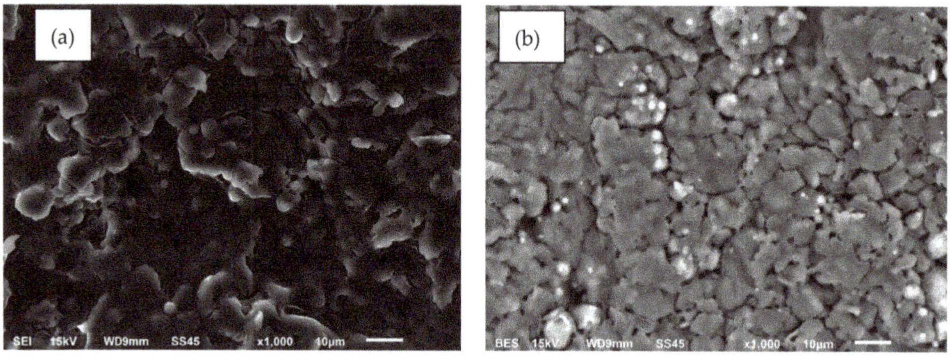

Figure 4. SEM microstructures of admixture. (**a**) SE mode; (**b**) BSE mode.

2.2. *Polypropylene Fibres*

Polypropylene fibres made from waste materials (plastic packaging) were used (Figure 5). The white polypropylene fibres (PPW) were made through a cutting process and their surface was modified into the extruder to increase its adhesion to the cementitious mix. Because of this modification process, the surface of PPW is irregular (Figures 6 and 7).

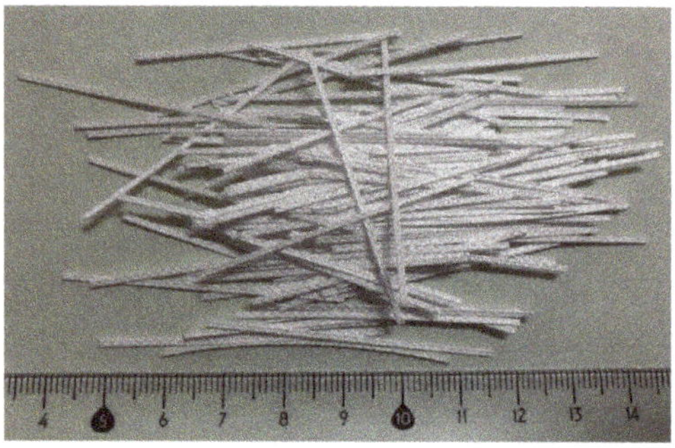

Figure 5. Polypropylene recycled fibres (PPW) with length ranging from 27.1 to 32.6 mm.

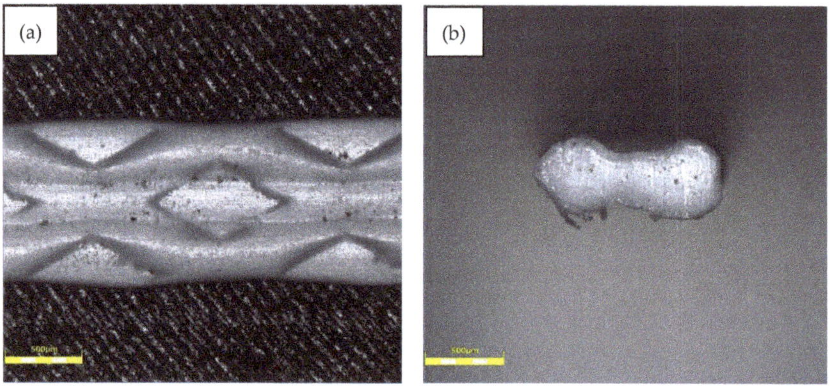

Figure 6. Light microscope images of the used fibres: (**a**) surface and (**b**) cross section.

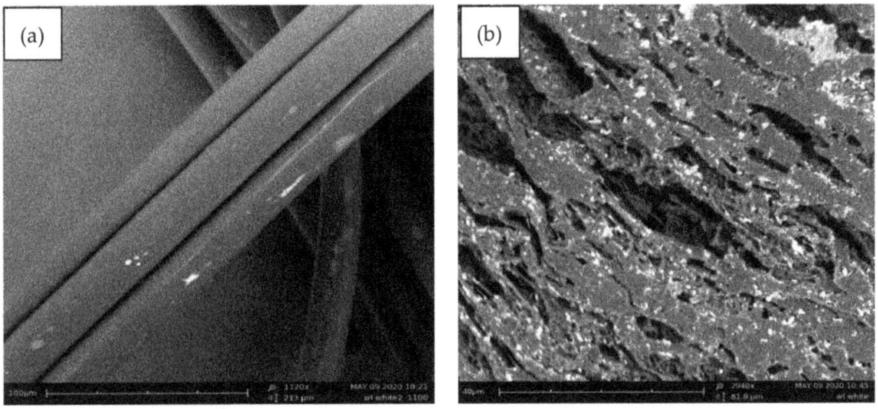

Figure 7. SEM images of the used fibres: (**a**) surface and (**b**) cross section.

Polypropylene fibres of 31.2 ± 0.5 mm in length and 1152.5 ± 10.0 μm in diameter were used. The average circumference was 538.2 ± 0.5 μm. The tensile strength of the fibres was about 520 MPa, the modulus of elasticity was about 7.5 GPa, and Poisson ratio was about 0.2. Fibre content ratios of 300, 600, 900, 1200, and 1500 g/m^3 were used.

2.3. Mix Composition

Six different modifications of concrete mixtures (reference—without fibres, and five with fibres) were produced. The polypropylene fibre content was about 0.0625%, 0.1250%, 0.1875%, 0.2500%, and 0.3125% of cement mass, respectively; see Table 3. A constant water to cement ratio, w_{eff}/c = 0.29, was used for all mixes, where w_{eff} was tap water content and c was the cement content. The admixture amount was 1.0% of the cement mass, which corresponds with [75].

Table 3. Mix proportions (1 m^3).

Mix Symbol	Cement [kg]	Water [kg]	Glass Powder [kg]	Glass Sand Aggregate		Fibre [g]
				0.1–0.9 mm	0.9–2.0 mm	
M0	480	140	600	510	790	0.0
M1						300
M2						600
M3						900
M4						1200
M5						1500

2.4. Mix Production

The technology of mixing individual components of the composite assumes using the "dry" and "wet" mixing methods. For the mixing stage, a high-speed planetary mixer (mixer) with variable speeds of stirrer rotation (three ranges of stirrer rotation speed) was used. Various mixing speeds and types of agitator were used. Two types of mixers were used: flat and "hook". The appropriate types of mixers were selected depending on the sequence of ingredients and their type. The following sequence and methods of mixing the composite components were proposed:

- Glass granulate of 0.9/1.5 mm fraction and a group of granulate fraction 0/0.9 mm—"dry" mixing, mixing time 2 min from the moment of filling the mixing container with the above-mentioned granulate fractions, mixing speed: gear 1, type of agitator: flat;

- Cement binder with filler (glass powder 0/200 μm)—"dry" mixing, mixing time: 2 min from the moment the mixing container is filled with the listed ingredients, mixing speed: gear 1, type of agitator: flat;
- Mixing water with liquid chemical admixture—mixing time: 2 min from the moment of adding both components to the measuring cylinder, mixing speed: gear 1 (mixing takes place with the measuring cylinder by means of a stirrer), type of stirrer: magnetic;
- Glass granulate with cement and glass powder—mixing time: 3 min from the moment of adding all ingredients to the mixing container of the mixer, mixing speed: gear 1, type of agitator: "hook", dry mixing;
- Glass granulate with cement and glass powder—mixing time: 3 min, mixing speed: gear 2, type of agitator: flat, mixing speed: gear 1, "dry" mixing;
- Glass granulate with cement and glass powder and mixing water with a liquid chemical admixture—mixing time: 3–4 min, mixing speed: gear 1, type of agitator: flat, "wet" mixing;
- Glass gargoyle with cement and glass powder, and mixing water with a liquid chemical admixture—mixing time: 3 min, mixing speed: gear 2, type of agitator: flat, "wet" mixing;
- Glass granulate with cement and glass powder and mixing water with a liquid chemical admixture—mixing time: 3 min, mixing speed: gear 3, type of agitator: flat, "wet" mixing.

3. Methodology

3.1. *Test on Mix*

Slump cone and air content were measured after production of mixture. The slump cone was investigated according to PNEN 12350-2:2019-07 standard [80]. In order to preserve the statistics of the results, measurements were made on five test samples for each mixture. In order to obtain the air content, the pressure method was investigated per ASTM C231 standard [81]. Five samples were used for each mixture.

3.2. *Test on Hardened Concrete*

After 28 days of curing, the hardened concrete was investigated. For this purpose, five samples were measured for each concrete mixture. In order to determine the mechanical properties of the manufactured concrete, ten specimens were used for each test.

3.2.1. Material Properties

The density of the prepared concrete samples (150 mm × 150 mm × 150 mm) was measured according to the standard PN-EN 12390-7:2011 [82].

3.2.2. Mechanical Properties

The mechanical properties of hardened concrete were measured using three methods: compressive strength, splitting strength, and flexural strength. For each test, the Zwick machine was used with a force range of 0–5000 kN (Zwick, Ulm, Germany). In addition, the modulus of elasticity and Poisson ratio were investigated.

1. *Static compression test of cubic samples*

The compressive strength test was carried out on cubic samples with dimensions of 150 mm × 150 mm × 150 mm according to the standard EN 12390-3:2019-07 [83] after 28 days of curing under standard conditions (21 °C, 50% humidity). Cubic samples, after being removed from the care bath, were dried of excess water. Compressive strength tests were carried out 30 min after the end of curing. The samples were placed in the vertical working space on the lower clamping plate.

1. *Static flexural test of beam samples*

A static bending test of beam samples was carried out in order to determine the flexural strength of the composite modified with waste fibres. Specimens with dimensions of 100 mm × 100 mm × 500 mm were prepared and subjected to a three-point flexural test on a testing machine according to the standard EN 12390-5:2019-08 [84]. The beams were fixed on supports with movable rollers and then loaded with concentrated force on the middle of the span. The spacing in the axes of the supports was set at 400 mm. The construction of supports and concentrated forces with movable rollers eliminated the negative effect of frictional forces on the course of testing beam samples. On the basis of the static three-point flexural test, the values of the maximum destructive force and the maximum destructive stresses were recorded.

1. *Static splitting test of cubic samples*

The splitting strength of the composite was determined by the splitting method according to EN 12390-6:2011 [85], the so-called "Brazilian" method. The test was carried out after 28 days of maintenance. The surfaces of the samples were cleaned of the sediment after the treatment was completed and the excess surface water was dried. The cubic composite samples were placed in a metal splitting test profile. The rate of stress increase in time was determined as 0.5 MPa/s. A static sample splitting test was performed. Based on the registered maximum failure force in the compression test, the specified value of failure stress was determined.

1. *Modulus of elasticity and Poisson coefficient*

The study of Young's modulus was performed with a non-destructive method per ASTM C215-19 standard [86] using the James TM E-Meter Mk II apparatus (Chicago, IL, USA), which uses principles based on determining the basic resonance frequencies of vibrations generated by the shocks measured by the accelerometer. This device tests three types of vibrations: longitudinal, transverse, and torsional. The method was applied on cylindrical samples with a diameter of 150 mm and a height of 300 mm after at least 28 days of hardening.

4. Results and Discussion

4.1. Fresh Properties

Table 4 presents the results of fresh mix properties. The average values of the five samples for each mix are given.

Table 4. Fresh properties.

Mix Symbol	Fibre Content [g/m^3]	Slump Cone [mm]	Air Content [%]
M0	0	58 ± 2	4.0 ± 0.5
M1	300	43 ± 2	3.8 ± 0.3
M2	600	35 ± 3	3.9 ± 0.2
M3	900	30 ± 3	3.5 ± 0.5
M4	1200	25 ± 3	3.6 ± 0.6
M5	1500	20 ± 3	3.6 ± 0.6

During the mixing, no process of agglomeration formation of fibres was observed for all cement–glass composite mixes, which is a main problem of concrete mixes with fibres. The fibres did not float to the surface nor did they sink to the bottom in the fresh mixes. They were mixed smoothly with the composite mixture. Figure 8 presents the results of the slump cone (SC) test. It can be observed that, by replacing the entire aggregate with glass aggregate and using glass powder as a filler [87], the reference mix and the mix with a lower fibre content were within slump class S2 [80]. Moreover, the slump cone obtained for the cement–glass composite was about 66% higher than for plain concrete with granite aggregate and a strength of 50 MPa [74]. An increase in slump with increasing fine glass aggregate was observed by Castro and Brito [88], while according to Limbachiya [89] and Taha [90], the use of glass sand resulted in the decrease in the workability of the concrete due to a lack of fine proportion.

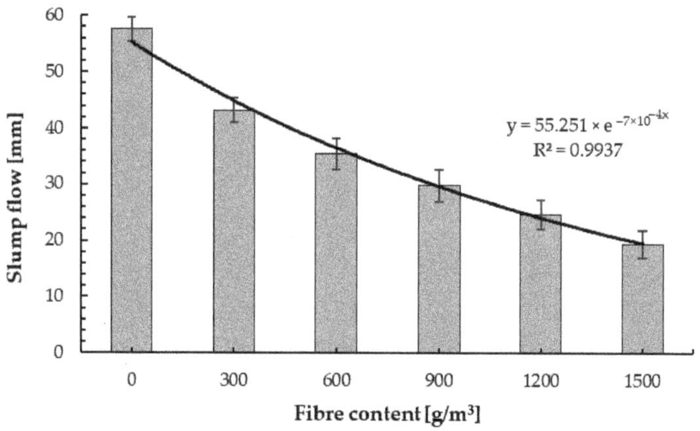

Figure 8. Slump test results.

The addition of PP fibres to the reference mix reduces its workability, which is analogous to observations for plain concrete reinforced PP fibres [44]. The mixes with 600 to 1500 g/m^3 of fibres were within slump class S1 [80]. Moreover, with the addition of 300, 600, 900, 1200, and 1500 g/m^3 of fibres, the decrease in slump was 25.9%, 39.7%, 48.3%, 56.9%, and 65.5%, respectively, compared with the reference specimen (without fibres). Practically, the same decrease in slump cone was observed for plain concrete reinforced with polypropylene fibres made from plastic packaging with the same contents [44]. A similar correlation of reduction in workability with an increase in fibre content was determined by other scientists [91,92].

The air content of the cement–glass composite was constant regardless of the PP fibre content and was equal to 4.0 ± 0.5% (Table 4). The addition of fibres did not affect the air content in the mixture. The obtained air content of the cement–glass composite, however, was two times higher than for plain concrete and for concrete with the addition of glass aggregate up to 20% of cement [74].

4.2. Hardened Properties

4.2.1. Density

The results of cement–glass composite density are given in Figure 9 and Table 5. The presented values are the average values of the five samples for each mix for density and ten samples for mechanical properties.

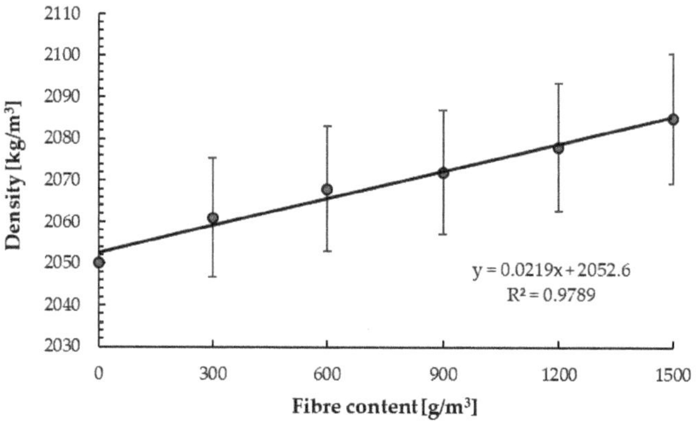

Figure 9. The correlation between the amount of fibres and the density of the cement–glass composite.

Table 5. The results of the properties of hardened samples.

Mix Symbol	Density [kg/m³]	Compressive Strength [MPa]		Flexural Strength [MPa]		Splitting Strength [MPa]		Elastic Modulus [GPa]	Poisson Ratio [[–]
		14 Day	28 Day	14 Day	28 Day	14 Day	28 Day		
M0	2050 ± 14	62.1 ± 0.8	82.5 ± 0.8	3.5 ± 0.1	4.9 ± 0.2	2.0 ± 0.7	2.5 ± 0.8	31 ± 1	0.15 ± 0.01
M1	2061 ± 14	63.8 ± 0.9	82.6 ± 0.5	3.7 ± 0.6	5.1 ± 0.1	2.7 ± 0.4	3.7 ± 0.4	32 ± 1	0.15 ± 0.02
M2	2068 ± 15	64.2 ± 0.3	83.0 ± 0.8	4.0 ± 0.3	5.3 ± 0.1	2.9 ± 0.9	4.0 ± 0.8	32 ± 1	0.15 ± 0.02
M3	2072 ± 15	66.8 ± 0.8	83.8 ± 0.3	4.2 ± 0.5	5.6 ± 0.1	4.3 ± 0.2	5.8 ± 0.3	32 ± 1	0.15 ± 0.01
M4	2078 ± 15	67.3 ± 0.3	84.4 ± 0.6	4.6 ± 0.3	5.9 ± 0.1	4.7 ± 0.1	6.4 ± 0.6	32 ± 2	0.15 ± 0.02
M5	2085 ± 16	67.6 ± 0.6	84.8 ± 0.6	4.9 ± 0.1	6.2 ± 0.2	5.7 ± 0.5	8.0 ± 0.6	32 ± 1	0.15 ± 0.01

With the increase in fibre content, the cement–glass composite density increased linearly (Figure 9), while the impact of glass sand addition was negligible. For the highest fibre content, the composite density increased by 1.7% compared with the base sample. This could be related to the composite production error or the occurring pozzolanic reaction [93]. Figure 10 presents the uniform distribution of glass aggregate and fibres in the sample.

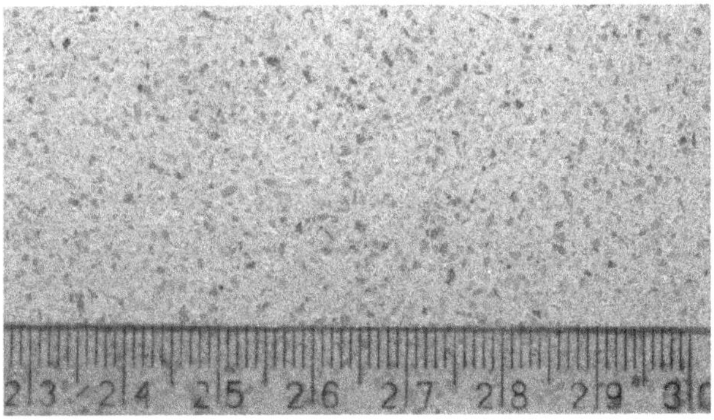

Figure 10. Distribution of components in the cement–glass composite.

The obtained cement–glass composite density was much lower than the density of plain concrete with granite aggregate (γ = 2205 ± 4 kg/m^3) and follows the demonstrated trend of the decrease in density with increasing glass fine aggregate [74]. Other scientists [94,95] observed the same results. Park et al. [95] obtained a linear decrease of concrete density with increasing waste glass aggregate content, while Lee et al. [94] determined that replacing the aggregate with glass sand up to the amount of 20–25% slightly affects concrete density.

4.2.2. Compressive Strength

The test results of compressive strength for samples of the cement–glass composite are shown in Figure 11. The compressive strength increased slightly with the increase in PP fibre addition, which was unexpected owing to the principle that fibres improve tensile strength, but not compressive strength [96–98]. Oni et al. [99] determined a slight increase for concrete with 0.3% PP fibre and a slight decrease with the addition of 0.4% of this fibre. For other tested types of polypropylene fibre, they obtained decrease in compressive strength. Moreover, Jiang et al. [100] reported 3.12% decrease in the compressive strength of PP fibre-reinforced concrete compared with the base sample.

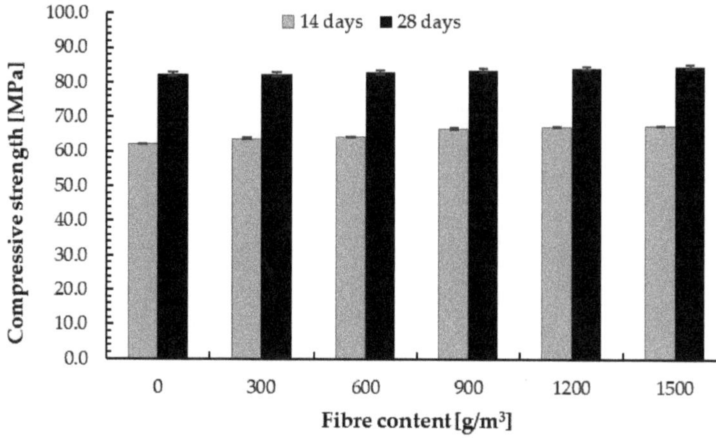

Figure 11. Compressive strength of the cement–glass composite depending on fibre content.

It can be observed that, after 14 days, 75% of the compressive strength after 28 days was obtained. Moreover, the compressive strength of the cement–glass composite was 1.5 times higher than for plain concrete of similar composition (cement CEM I 52.5R, granite aggregate 0/4 mm, and with the addition of a superplasticizer to reduce the amount of water to w/c = 0.26), for which the compressive strength was f_c = 53 MPa [74]. This is probably the result of the use of glass powder to ensure the continuity of the internal structure of the material, which in turn results in an increase in mechanical strength by reducing the air pores in the cement hydration process. In addition, according to observations of other scientists, the pozzolanic reactivity of fine waste glass with a particle size below 100 μm is observed as an increase in compressive strength [49–51] due to pozzolanic reaction. Moreover, this is in line with the observation that compressive strength improves with increase in the content of glass fine aggregate [74,101,102]. Lee et al. [94] obtained a 34.3% increase in compressive strength compared with plain concrete for concrete with fine glass aggregate with a particle size of 0–0.6 mm. Chung et al. [103] also reported that the use of aggregates with size less than 4.0 mm makes it possible to achieve an improvement in compressive strength. Using glass powder in concrete mixture could result in higher compressive strength, as increases in the compressive strength of concrete with glass powder have been found in other studies. Bajad et al. [104] determined that compressive strength increases with the addition of glass powder as a cement replacement at a ratio of up to 20%, and then decreases. Improvements in the long-term compressive strength of concrete containing fine glass powder have been reported by other scientists [36,105]. The increase in compressive strength could be caused by the pozzolanic reaction of very fine particles [57,90,106,107]. According to Shi et al. [108] the strength activity indices of fine glass powder with a size of 40–700 μm were 70% to 74% at 7 and 28 days, respectively. Kamali and Ghahremaninezhad [109] and Ling and Poon [66] reported that smaller particle sizes of aggregate enhance the aggregate–cement matrix bonding strength. Yamada et al. [110] demonstrated the critical particle size to range from 0.15 to 0.30 mm for the occurring pozzolanic reaction, while Jin et al. [111], Idir et al. [112], and Xie et al. [113] determined this size to be from 0.60 to 1.18 mm.

4.2.3. Flexural Strength

The results of the flexural test of the cement–glass composite are shown in Table 5 and Figure 12. It can be observed that, with the addition of 300, 600, 900, 1200, and 1500 g/m^3 of PP fibre, flexural strength increased compared with the base sample by 4.1%, 8.2%, 14.3%, 20.4%, and 26.5%, respectively. These values were obtained for a very small amount of fibre ranging from 0.0625% to 0.3125% of the cement weight. Thus, the addition of polypropylene fibre improves the flexural strength of the cement–glass composite. This is analogous to fibre-reinforced concrete with natural aggregate [95,114,115]. Nili and Afroughdaset [116] obtained a 22% improvement for concrete with silica fume and 0.3% of PP fibre. Satisha et al. [117] also determined about a 30% increase in flexural strength for concrete with 2.0% addition of PP fibres. About a 37% increase in flexural strength for concrete with 1.0 wt.% polypropylene fibres was obtained by Badogiannis et al. [118]. Other scientists reported an increase in flexural strength with addition of PP fibre ranging from about 10% [119–121] to 35% [118]. A higher increase in flexural strength was observed in the literature, but for significant proportions of PP fibres [121–123].

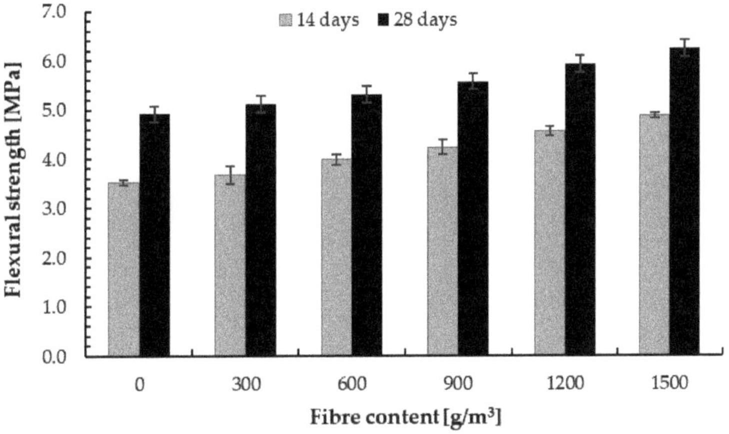

Figure 12. Flexural strength of the cement–glass composite depending on fibre content.

Moreover, after 14 days of curing, flexural strength ranging from about 70% to 80% of the target flexural strength was obtained. The increase in 14-day flexural strength was 5.7%, 14.3%, 20.0%, 31.4%, and 40.0%, respectively, compared with the base sample. This proves the large use of recycled fibres and is a continuation of the research presented by Malek et al. [44].

The flexural strength of the cement–glass composite was almost half that of the flexural strength of plain concrete of a similar composition, but with granite aggregate (f_{tk} = 10.5 ± 0.3 MPa [74]). According to Tan and Du [112], the reduction in flexural strength is caused by a decrease in adhesive strength at the glass particle surface and cement matrix and, additionally, micro-cracks in the case of clear glass aggregate. The effect of the weaker bonding of glass aggregate with the cement matrix is more important in the flexural test than in the compression test. Moreover, this runs contrary to the observation that bending strength is enhanced with the addition of fine glass aggregate, and thus the increase in fracture toughness [24,74]. The reduction in flexural strength, however, was demonstrated by other scientists [103]. Ling and Poon [66] obtained about a 30% decrease in flexural strength for concrete with 100% glass aggregate (60 wt.% glass aggregate size from 0 to 2.36 mm and 40 wt.% size from 2.36 to 5.00 mm).

4.2.4. Splitting Strength

Table 5 presents the splitting strength for the cement–glass composite with different fibre content. With the increase in PP fibre content, the splitting strength increased linearly; see Figure 13. For the cement–glass composition with 300, 600, 900, 1200, and 1500 g/m^3 of PP fibre, the 28-day splitting strength was 35%, 45%, 115%, 135%, and 185% higher, respectively, than for the reference sample, while the increase in the 14-day splitting strength was 48%, 68%, 132%, 156%, and 220%, respectively, compared with the base sample. This significant improvement was obtained for a very small amount of fibre ranging from

0.0625% to 0.3125% of the cement weight. Thus, the addition of polypropylene fibre improves the splitting strength of the cement–glass composite. The same phenomenon was observed by other scientists for concrete with natural aggregate [95,124,125]. The fibres are able to bridge the cracks and transfer stress across the cracks [126,127]. The fibre-reinforced composite is destroyed when the fibre slides out of the matrix or breaks (in the second case, the load is redistributed to the other fibers [128]). Thus, the method of damage of the fibre-reinforced cement–glass composite mostly depends on the strength of the materials and the adhesion of the fibers to the matrix [129–131].

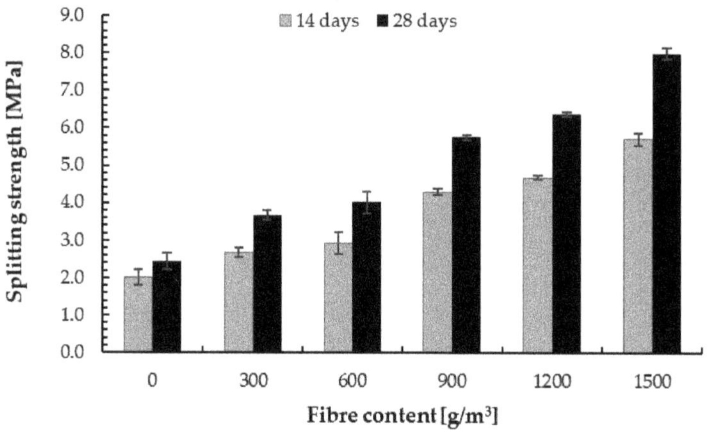

Figure 13. Splitting strength of the cement–glass composite depending on fibre content.

The increase in splitting strength was much larger than the increase in flexural strength. Analogous to flexural strength, however, after 14 days of curing, splitting strength from about 70% to 80% of the target flexural strength was obtained.

The splitting strength of the cement–glass composite was about two times lower than for plain concrete of a similar composition, but with granite aggregate (f_r = 4.12 MPa [74]). With the addition of glass aggregate, the splitting strength improved compared with the base sample [7,74]. The sample with 100 wt.% of glass sand aggregate, however, was lower. This may indicate that the strength increases with increasing cullet content and then decreases.

4.2.5. Modulus of Elasticity and Poisson Ratio

The results of the modulus of elasticity are shown in Table 5. In this study, the modulus of elasticity obtained was about 31 ± 1 GPa regardless of fibre content, and was equal to normal concrete with granite aggregate (E = 32 ± 1 GPa [74]). According to other papers [74,98], an insignificant effect of glass aggregate content on the elastic modulus can be observed.

The addition of PP fibre up to 0.3125% of the cement did not affect the Poisson ratio (Table 5).

5. Conclusions

The purpose of the research was to assess the possibility of using a large amount of glass cullet as a substitute for concrete components. Glass powder as filler and 100% of glass aggregate were used. The cement–glass composite exhibited low tensile strength and brittle failure. In order to improve tensile strength, the effects of adding polypropylene fibres on the mechanical properties of the composite were examined. The polypropylene fibre content was 0.0625%, 0.1250%, 0.1875%, 0.2500%, and 0.3125% of cement mass, respectively. Based on the results of this experimental investigation, the following key conclusions can be drawn:

- An effect of a decrease in the slump cone with the addition of PP fibres was noted; the reference mix and the mix with a lower fibre content were within slump class S2, but the mix with higher PP fibre content was within slump class S1.
- The amount of air in the cement–glass composite mix was equal to 4.0 ± 0.5%. The addition of fibres did not affect the air content of the mixture.
- With the increase of PP fibre content, the density of the cement–glass composite increased, but this effect was negligible (2–3% compared with the reference sample).
- With the addition of 0.0625%, 0.1250%, 0.1875%, 0.2500%, and 0.3125 wt.% polypropylene fibre, the increase in flexural strength of the cement–glass composite compared with the reference sample was about 4%, 8%, 14%, 20%, and 27%, respectively, while the increase in splitting strength was about 48%, 60%, 132%, 156%, and 220%, respectively. The effect of the increase in splitting strength was much larger than the increase in flexural strength. The compressive strength increased slightly with the increasing PP fibre content, which was unexpected owing to the principle that fibres improve tensile strength, but not compressive strength (0.1%, 0.6%, 1.6%, 2.3%, and 2.8% increase for 0.025, 0.050, 0.075, 0.100, and 0.125 wt.% polypropylene fibre, respectively).
- The elastic modulus of the cement–glass composite with content of 0.0625%, 0.1250%, 0.1875%, 0.2500%, and 0.3125 wt.% PP fibre was about 31 ± 1 GPa regardless of fibre content, and was equal to plain concrete with granite aggregate.
- The addition of PP fibre up to 0.3125% of the cement did not affect the Poisson ratio.

High values of flexural and splitting strength are the results of polypropylene fibres. This research will be subject to further testing. Other types of cement, glass waste and its mixes, and different contents of glass powder, with particular emphasis on long-term fatigue tests, are planned in this respect. In addition, it should be emphasized that the modulus of elasticity of the tested concrete composite is very low, which may result in greater deflection of the structure. This is why elements with no significant deflections, such as columns or sheet piling, can be made of a cement–glass composite (preferably with the highest obtained mechanical properties, e.g., recipe M5 with addition of 0.3125% PP fibres).

Author Contributions: Conceptualisation, M.M. and M.K.; Data Curation, M.K. and M.M.; Formal analysis, M.K.; Funding Acquisition, M.M.; Investigation, M.M., W.Ł., J.K., and D.D.; Methodology, M.M. and M.K.; Project Administration, M.M.; Resources, M.M. and W.Ł.; Supervision, M.K.; Validation, M.K.; Visualisation, M.K. and M.M.; Writing—Original Draft Preparation, M.K. and M.M.; Writing—Review & Editing, M.K. All authors have read and agreed to the published version of the manuscript

Funding: This research received no external funding besides statutory research of particular scientific units.

Institutional Review Board Statement: Not applicable.

Informed Consent Statement: Not applicable.

Data Availability Statement: Data is contained within the article.

Acknowledgments: This work was financially supported by the Dean of Faculty of Civil Engineering and Geodesy of the Military University of Technology as part of scholarship no. 1/DHP/2020.

Conflicts of Interest: The authors declare no conflict of interest.

References

1. Kijeński, J.; Błędzki, A.; Jeziórska, R. *Recovery and Recycling of Polymeric Materials [Odzysk i Recykling Materiałów Polimerowych]*; PWN: Warsaw, Poland, 2011.
2. Plastics Europe. *The Compelling Facts about Plastics–Analysis of Plastics Production, Demand and Recovery for 2008 in Europe*; Plastics Europe: Brussels, Belgium, 2008.
3. Knap, P. The Influence of Raw Materials on the Load-Bearing Capacity of Plastic Connectors in Building Fastenings [Wpływ Surowców na Nośność Łączników Tworzywowych w Zamocowaniach Budowlanych]. Ph.D. Thesis, Building Research Institute, Varsav, Poland, 20 April 2017.
4. Plastics Europe Polska. *Analysis of Production, Demand and Recovery of Plastics in Europe in 2011 [Analiza Produkcji, Zapotrzebowanie oraz Odzysk Tworzyw Sztucznych w Europie w Roku 2011]*; Plastics Europe Polska: Warszawa, Poland, 2012.

5. Pilz, H.; Brandt, B.; Fehringer, R. *Wpływ Tworzyw Sztucznych na Zużycie Energii Oraz na Emisję Gazów Cieplarnianych w Europie z Uwzględnieniem Całego Cyklu Życia Wyrobów*; Denkstatt GmbH: Vienna, Austria, 2010.
6. Baker Institute. Pandemic, Plastics and the Continuing Quest for Sustainability. *Forbes Media LLC 2020*. Available online: https://www.forbes.com/sites/thebakersinstitute/2020/04/14/pandemic-plastics-and-the-continuing-quest-for-sustainability/#36f2b4a477b4 (accessed on 27 November 2020).
7. Tan, K.H.; Du, H. Use of waste glass as sand in mortar: Part I—Fresh, mechanical and durability properties. *Cem. Concr. Compos.* **2013**, *35*, 109–117. [CrossRef]
8. EPA. United States Environmental Protection Agency, Advancing Sustainable Materials Management: 2017 Fact Sheet, EPA. United States Environmental Protection Agency. 2017. Available online: https://www.epa.gov/sites/production/files/2019-11/documents/2017_facts_and_figures_fact_sheet_final.pdf (accessed on 27 November 2020).
9. LCQ9. Recovery and Recycling of Waste Glass. 2020. Available online: https://www.info.gov.hk/gia/general/202001/08/P2020010800455p.htm (accessed on 27 November 2020).
10. Bostanci, S.C.; Limbachiya, M.; Kew, H. Portland-composite and composite cement concretes made with coarse recycled and recycled glass sand aggregates: Engineering and durability properties. *Constr. Build. Mater.* **2016**, *128*, 324–340. [CrossRef]
11. Liu, F.; Ding, W.; Qiao, Y. An experimental investigation on the integral waterproofing capacity of polypropylene fiber concrete with fly ash and slag powder. *Constr. Build. Mater.* **2019**, *212*, 675–686. [CrossRef]
12. Sabet, F.A.; Libre, N.A.; Shekarchi, M. Mechanical and durability properties of self consolidating high performance concrete incorporating natural zeolite, silica fume and fly ash. *Constr. Build. Mater.* **2013**, *44*, 175–184. [CrossRef]
13. Rudnicki, T.; Wołoszka, P. The use of technology whitetopping in the aspect of implementation of repairs of flexible pavements. *Bull. Mil. Univ. Technol.* **2016**, *65*, 3.
14. Limbachiya, M.; Meddah, M.S.; Ouchagour, Y. Use of recycled concrete aggregate in fly-ash concrete. *Constr. Build. Mater.* **2012**, *27*, 439–449. [CrossRef]
15. Gesoğlu, M.; Güneyisi, E.; Özbay, E. Properties of self-compacting concretes made with binary, ternary, and quaternary cementitious blends of fly ash, blast furnace slag, and silica fume. *Constr. Build. Mater.* **2009**, *23*, 1847–1854. [CrossRef]
16. Kadela, M.; Kukiełka, A. Influence of foaming agent content in fresh concrete on elasticity modulus of hard foam concrete. In *Brittle Matrix Composites 11, Proceedings of the 11th International Symposium on Brittle Matrix Composites BMC 2015, Warsaw, Poland, 28–30 September 2015*; Institute of Fundamental Technological Research PAS: Warsaw, Poland, 2015; pp. 489–496.
17. Rudnicki, T.; Jurczak, R. Recycling of a Concrete Pavement after over 80 Years in Service. *Materials* **2020**, *13*, 2262. [CrossRef]
18. Osborne, G.J. Durability of Portland blast-furnace slag cement concrete. *Cem. Concr. Compos.* **1999**, *21*, 11–21. [CrossRef]
19. De Domenico, D.; Faleschini, F.; Pellegrino, C.; Ricciardi, G. Structural behavior of RC beams containing EAF slag as recycled aggregate: Numerical versus experimental results. *Constr. Build. Mater.* **2018**, *171*, 321–337. [CrossRef]
20. Menéndez, G.; Bonavetti, V.; Irassar, E.F. Strength development of ternary blended cement with limestone filler and blast-furnace slag. *Cem. Concr. Compos.* **2003**, *25*, 61–67. [CrossRef]
21. Hama, S.M.; Hilal, N.N. Fresh properties of self-compacting concrete with plastic waste as partial replacement of sand. *Int. J. Sustain. Built Environ.* **2017**, *6*, 299–308. [CrossRef]
22. Choi, Y.W.; Moon, D.J.; Chumg, J.S.; Cho, S.K. Effects of waste PET bottles aggregate on the properties of concrete. *Cem. Concr. Compos.* **2005**, *35*, 776–781. [CrossRef]
23. Saxena, R.; Siddique, S.; Gupta, T.; Sharma, R.K.; Chaudhary, S. Impact resistance and energy absorption capacity of concrete containing plastic waste. *Constr. Build. Mater.* **2018**, *176*, 415–421. [CrossRef]
24. Mohammadinia, A.; Wong, Y.C.; Arulrajah, A.; Horpibulsuk, S. Strength evaluation of utilizing recycled plastic waste and recycled crushed glass in concrete footpaths. *Constr. Build. Mater.* **2019**, *197*, 489–496. [CrossRef]
25. Choi, Y.W.; Moon, D.J.; Kim, Y.J.; Lachemi, M. Characteristics of mortar and concrete containing fine aggregate manufactured from recycled waste polyethylene terephthalate bottles. *Constr. Build. Mater.* **2009**, *23*, 2829–2835. [CrossRef]
26. Gesoğlu, M.; Güneyisi, E.; Hansu, O.; Etli, S.; Alhassan, M. Mechanical and fracture characteristics of self-compacting concretes containing different percentage of plastic waste powder. *Constr. Build. Mater.* **2017**, *140*, 562–569. [CrossRef]
27. Asokan, P.; Osmani, M.; Price, A.D.F. Improvement of the mechanical properties of glass fibre reinforced plastic waste powder filled concrete. *Constr. Build. Mater.* **2010**, *24*, 448–460. [CrossRef]
28. Jackowski, M.; Małek, M.; Życiński, W.; Łasica, W.; Owczarek, M. Characterization of new recycled polymers shots addition for the mechanical strength of concrete. *Mater. Tehnol.* **2020**, *54*, 355–358. [CrossRef]
29. Mahdi, F.; Khan, A.A.; Abbas, H. Physiochemical properties of polymer mortar composites using resins derived from post-consumer PET bottles. *Cem. Concr. Compos.* **2007**, *29*, 241–248. [CrossRef]
30. Pacheco-Torgal, F.; Ding, Y.; Jalali, S. Properties and durability of concrete containing polymeric wastes (tyre rubber and polyethylene terephthalate bottles): An overview. *Constr. Build. Mater.* **2012**, *30*, 714–724. [CrossRef]
31. Yin, S.; Tuladhar, R.; Shi, F.; Combe, M.; Collister, T.; Sivakugan, N. Use of macro plastic fibers in concrete: A review. *Constr. Build. Mater.* **2015**, *93*, 180–188. [CrossRef]
32. Kou, S.C.; Lee, G.; Poon, C.S.; Lai, W.L. Properties of lightweight aggregate concrete prepared with PVC granules derived from scraped PVC pipes. *Waste Manag.* **2009**, *29*, 621–628. [CrossRef] [PubMed]
33. Pesic, N.; Zivanovic, S.; Gasrcia, R.; Papastergiou, P. Mechanical properties of concrete reinforced with recycled HDPE plastic fibres. *Constr. Build. Mater.* **2016**, *115*, 362–370. [CrossRef]

34. Panyakapo, P.; Panyakapo, M. Reuse of thermosetting plastic waste for lightweight concrete. *Waste Manag.* **2008**, *28*, 1581–1588. [CrossRef]
35. Kan, A.; Demirboga, R. A new technique of processing for waste-expanded polystyrene foams as aggregates. *J. Mater. Proc. Technol.* **2009**, *209*, 2994–3000. [CrossRef]
36. Kim, S.B.; Yi, N.H.; Kim, H.Y.; Kim, J.H.J.; Song, Y.C. Material and structural performance evaluation of recycled PET fiber reinforced concrete. *Cem. Concr. Res.* **2010**, *32*, 232–240. [CrossRef]
37. Szcześniak, A.; Stolarski, A. Dynamic Relaxation Method for Load Capacity Analysis of Reinforced Concrete Elements. *Appl. Sci.* **2018**, *8*, 396. [CrossRef]
38. Kozłowski, M.; Kadela, M. Mechanical Characterization of Lightweight Foamed Concrete. *Adv. Mater. Sci. Eng.* **2018**, *2018*, 1–8. [CrossRef]
39. Kan, A.; Demirboga, R. A novel material for lightweight concrete production. *Cem. Concr. Compos.* **2009**, *31*, 489–495. [CrossRef]
40. Kadela, M.; Kukiełka, A.; Małek, M. Characteristics of Lightweight Concrete Based on a Synthetic Polymer Foaming Agent. *Materials* **2020**, *13*, 4979. [CrossRef] [PubMed]
41. Garbacz, A.; Szmigiera, E.D.; Protchenko, K.; Urbański, M. On Mechanical Characteristics of HFRP Bars with Various Types of Hybridization. In *International Congress on Polymers in Concrete (ICPIC 2018): Polymers for Resilient and Sustainable Concrete Infrastructure*; Taha, M.M.R., Girum, U., Moneeb, G., Eds.; Springer: Berlin/Heidelberg, Germany, 2018; pp. 653–658.
42. Dudek, D.; Kadela, M. Pull-Out Strength of Resin Anchors in Non-cracked and Cracked Concrete and Masonry Substrates. *Procedia Eng.* **2016**, *161*, 864–867. [CrossRef]
43. Knap, P.; Dudek, D. Impact of the Degree of Concrete Cracking on the Pull-out Resistance of Steel and Plastic/Metal Sleeve Anchors. *IOP Conf. Series: Mater. Sci. Eng.* **2017**, *245*, 022091. [CrossRef]
44. Małek, M.; Jackowski, M.; Łasica, W.; Kadela, M. Characteristics of Recycled Polypropylene Fibers as an Addition to Concrete Fabrication Based on Portland Cement. *Materials* **2020**, *13*, 1827. [CrossRef] [PubMed]
45. Meyer, C. The greening of the concrete industry. *Cem. Concr. Compos.* **2009**, *31*, 601–605. [CrossRef]
46. Nassar, R.-U.-D.; Soroushian, P. Green and durable mortar produced with milled waste glass. *Mag. Concr. Res.* **2012**, *64*, 605–615. [CrossRef]
47. Nassar, R.-U.-D.; Soroushian, P. Strength and durability of recycled aggregate concrete containing milled glass as partial replacement for cement. *Constr. Build. Mater.* **2012**, *29*, 368–377. [CrossRef]
48. Ghaffary, A.; Moustafa, M.A. Synthesis of Repair Materials and Methods for Reinforced Concrete and Prestressed Bridge Girders. *Materials* **2020**, *13*, 4079. [CrossRef]
49. Ramdani, S.; Guettala, A.; Benmalek, M.L.; Aguiar, J.B. Physical and mechanical performance of concrete made with waste rubber aggregate, glass powder and silica sand powder. *J. Build. Eng.* **2019**, *21*, 302–311. [CrossRef]
50. Ankur, M.; Randheer, S. Comparative Study of Waste Glass Powder as Pozzolanic Material in concrete. Bachelor's Thesis, Department of Civil Engineering, National Institute of Technology, Deemed University, Rourkela, India, 18, May, 2012.
51. Kou, S.C.; Xing, F. The effect of recycled glass powder and reject fly ash on the mechanical properties of fiber-reinforced ultralight performance concrete. *Adv. Mater. Sci. Eng.* **2012**, *2012*, 263243. [CrossRef]
52. Hendi, A.; Mostofinejad, D.; Sedaghatdoost, A.; Zohrabi, M.; Naeiimi, N.; Tavakolinia, A. Mix design of the green self-consolidating concrete: Incorporating the waste glass powder. *Constr. Build. Mater.* **2019**, *199*, 369–384. [CrossRef]
53. Soliman, N.A.; Tagnit-Hamou, A. Development of ultra-high-performance concrete using glass powder—Towards ecofriendly concrete. *Constr. Build. Mater.* **2016**, *125*, 600–612. [CrossRef]
54. Kadela, M.; Kozłowski, M.; Kukiełka, A. Application of foamed concrete in road pavement–weak soil system. *Procedia Eng.* **2017**, *193*, 439–446. [CrossRef]
55. Spiesz, P.; Rouvas, S.; Brouwers, H.J.H. Utilization of waste glass in translucent and photocatalytic concrete. *Const. Build. Mat.* **2016**, *128*, 436–448. [CrossRef]
56. Najad, A.A.A.; Kareem, H.J.K.; Azline, N.; Ostovar, N. Waste glass as partial replacement in cement—A review. *IOP Conf. Ser.: Earth Environ. Sci.* **2019**, *357*, 012023. [CrossRef]
57. Federico, L.M.; Chidiac, S.E. Waste glass as a supplementary cementitious material in concrete—Critical review of treatment methods. *Cem. Concr. Compos.* **2009**, *31*, 606–610. [CrossRef]
58. Mirzahosseini, M.; Riding, K.A. Effect of curing temperature and glass type on the pozzolanic reactivity of glass powder. *Cem. Concr. Res.* **2014**, *58*, 103–111. [CrossRef]
59. Johnson, C.D. Waste glass as coarse aggregate for concrete. *J. Test. Eval.* **1974**, *2*, 344–350.
60. Topçu, İ.B.; Canbaz, M. Properties of concrete containing waste glass. *Cem. Concr. Res.* **2004**, *34*, 267–274. [CrossRef]
61. Ling, T.-C.; Poon, C.-S. Utilization of recycled glass derived from cathode ray tube glass as fine aggregate in cement mortar. *J. Hazard Mater.* **2011**, *192*, 451–456. [CrossRef]
62. Yu, X.; Tao, Z.; Song, T.Y.; Pan, Z. Performance of concrete made with steel slag and waste glass. *Constr. Build. Mater.* **2016**, *114*, 737–746. [CrossRef]
63. Limbachiya, M.; Meddah, M.S.; Fotiadou, S. Performance of granulated foam glass concrete. *Constr. Build. Mater.* **2012**, *28*, 759–768. [CrossRef]
64. Tittarelli, F.; Giosuè, C.; Mobili, A. Recycled Glass as Aggregate for Architectural Mortars. *Int. J. Concr. Struct. Mater.* **2018**, *12*, 1–11. [CrossRef]

65. Kou, S.C.; Poon, C.S. Properties of self-compacting concrete prepared with recycled glass aggregate. *Cem. Concr. Compos.* **2009**, *31*, 107–113. [CrossRef]
66. Ling, T.-C.; Poon, C.-S. Properties of architectural mortar prepared with recycled glass with different particle sizes. *Mater. Des.* **2011**, *32*, 2675–2684. [CrossRef]
67. Yousefi, A.; Tang, W.; Khavarian, M.; Fang, C.; Wang, S. Thermal and Mechanical Properties of Cement Mortar Composite Containing Recycled Expanded Glass Aggregate and Nano Titanium Dioxide. *Appl. Sci.* **2020**, *10*, 2246. [CrossRef]
68. EEA. *EEA–NEC Report, NEC Directive Status Report 2008*; European Environment Agency: København, Denmark, 2008.
69. EPA (Environmental Protection Agency). *Available and Emerging Technologies for Reducing Greenhouse Gas Emissions from the Portland Cement Industry*; EPA: Washington, DC, USA, 2010.
70. USGS (US Geological Survey). *Background Facts and Issues Concerning Cement and Cement Data*; USGS: Reston, VA, USA, 2005.
71. Małek, M.; Jackowski, M.; Życiński, W.; Wachowski, M. Characterization of new fillers addition on mechanical strength of concrete. *Mater. Tehnol.* **2019**, *53*, 239–243. [CrossRef]
72. Rudnicki, T. The method of aggregate skeleton in self compacting concrete designing with segment regression. *CWB-1* **2016**, *1*, 10–19.
73. Szcześniak, A.; Zychowicz, J.; Stolarski, A. Influence of Fly Ash Additive on the Properties of Concrete with Slag Cement. *Materials* **2020**, *13*, 3265. [CrossRef]
74. Łasica, W.; Małek, M.; Szcześniak, Z.; Owczarek, M. Characterization of recycled glass-cement composite: Mechanical strength. *Mater. Technol.* **2020**, *54*, 473–477. [CrossRef]
75. Małek, M.; Łasica, W.; Jackowski, M.; Kadela, M. Effect of waste glass addition as replacement of fine aggregate on properties of concrete. *Materials* **2020**, *13*, 3189. [CrossRef] [PubMed]
76. PN EN 196-6:2011. *Methods of Testing Cement—Part 6: Determination of Fineness*; European Committee for Standardization: Brussels, Belgium, 2019.
77. PN EN 196-1:2016-07. *Methods of Testing Cement—Part 1: Determination of Strength*; European Committee for Standardization: Brussels, Belgium, 2016.
78. Górażdże Group: Cement, Concrete, Aggregate. Technical Data Sheet CEM I 42.5 R. Available online: http://www.gorazdze.pl (accessed on 27 November 2020).
79. E-grit Polska. Innovative Product for Cleaning the Surface PANGLASS, Safety Data Sheet Panglass-Glass Flour, Glass Granules. Available online: https://egrit.pl/ (accessed on 27 November 2020).
80. PN EN 12350-2:2019-07. *Testing Fresh Concrete—Part 2: Slump Test*; European Committee for Standardization: Brussels, Belgium, 2019.
81. ASTM C231/C231M-17a. *Standard Test Method for Air Content of Freshly Mixed Concrete by the Pressure Method*; ASTM International: West Conshohocken, PA, USA, 2017; Available online: www.astm.org (accessed on 15 April 2017).
82. EN 12350-7:2019-08. *Testing Fresh Concrete—Part 7: Air Content-Pressure Method*; European Committee for Standardization: Brussels, Belgium, 2019.
83. EN 12390-3:2019-07. *Testing Hardened Concrete—Part 3: Compressive Strength of Test Specimens*; European Committee for Standardization: Brussels, Belgium, 2019.
84. EN 12390-5:2019-08. *Testing Hardened Concrete—Part 5: Flexural Strength of Test Specimens*; European Committee for Standardization: Brussels, Belgium, 2019.
85. EN 12390-6:2011. *Testing Hardened Concrete—Part 6: Tensile Splitting Strength of Test Specimens*; European Committee for Standardization: Brussels, Belgium, 2011.
86. ASTM C215-19. *Standard Test Method for Fundamental Transverse, Longitudinal, and Torsional Resonant Frequencies of Concrete Specimens*; ASTM International: West Conshohocken, PA, USA, 2019; Available online: www.astm.org (accessed on 1 December 2019).
87. Moosberg-Bustnes, H.; Lagerblad, B.; Forssberg, E. The function of fillers in concrete. *Mater. Struct.* **2004**, *37*, 74. [CrossRef]
88. Castro, S.; Brito, J. Evaluation of the durability of concrete made with crushed glass aggregates. *J. Clean. Prod.* **2013**, *41*, 7–14. [CrossRef]
89. Limbachiya, M.C. Bulk engineering and durability properties of washed glass sand concrete. *Constr. Build. Mater.* **2009**, *23*, 1078–1083. [CrossRef]
90. Taha, B.; Nounu, G. Utilizing waste recycled glass as sand/cement replacement in concrete. *J. Mater. Civ. Eng.* **2009**, *21*, 709–721. [CrossRef]
91. Mohammadhosseini, H.; Tahir, M.M. Durability performance of concrete incorporating waste metalized plastic fibres and palm oil fuel ash. *Constr. Build. Mater.* **2008**, *180*, 92–102. [CrossRef]
92. Bayasi, Z.; Zeng, J. Properties of polypropylene fiber reinforced concrete. *ACI Mater. J.* **1993**, *90*, 605–610.
93. Du, H.; Tan, K.H. Use of waste glass as sand in mortar: Part II—Alkali-silica reaction and migration methods. *Cem. Con. Compos.* **2013**, *35*, 118–126. [CrossRef]
94. Lee, G.; Poon, C.S.; Wong, Y.L.; Ling, T.C. Effects of recycled fine glass aggregates on the properties of dry–mixed concrete blocks. *Constr. Build. Mater.* **2013**, *38*, 638–643. [CrossRef]
95. Park, S.B.; Lee, B.C.; Kim, J.H. Studies on mechanical properties of concrete containing waste glass aggregate. *Cem. Concr. Res.* **2004**, *34*, 2181–2189. [CrossRef]
96. Alhozaimy, A.M.; Soroushian, P.; Mirza, F. Mechanical properties of polypropylene fiber reinforced concrete and the effects of pozzolanic materials. *Cem. Concr. Compos.* **1996**, *18*, 85–92. [CrossRef]

97. Aulia, T.B. Effects of polypropylene fibres on the properties of high-strength concretes. *LACER* **2002**, *7*, 43–59.
98. Richardson, A. Compressive strength of concrete with polypropylene fibre additions. *Struc. Surv.* **2006**, *24*, 138–153. [CrossRef]
99. Oni, B.; Xia, J.; Liu, M. Mechanical properties of pressure moulded fibre reinforced pervious concrete pavement brick. *Case. Stud. Constr. Mater.* **2020**, *13*, e00431. [CrossRef]
100. Jiang, C.; Fan, K.; Wu, F.; Chen, D. Experimental study on the mechanical properties and microstructure of chopped basalt fibre reinforced concrete. *Mater. Des.* **2014**, *58*, 187–193. [CrossRef]
101. Grujoska, V.; Grujoska, J.; Samardzioska, T.; Jovanoska, M. Waste glass effects on fresh and hardened concrete. In Proceedings of the 7th International Conference "Civil Engineering—Science and Practice", Kolasin, Montenegro, 10–14 March 2020.
102. Jurczak, R.; Szmatuła, F.; Rudnicki, T.; Korentz, J. Effect of Ground Waste Glass Addition on the Strength and Durability of Low Strength Concrete Mixes. *Materials* **2021**, *14*, 190. [CrossRef]
103. Chung, S.-Y.; Elrahman, M.A.; Sikora, P.; Rucinska, T.; Horszczaruk, E.; Stephan, D. Evaluation of the Effects of Crushed and Expanded Waste Glass Aggregates on the Material Properties of Lightweight Concrete Using Image-Based Approaches. *Materials* **2017**, *10*, 1354. [CrossRef]
104. Bajad, M.N.; Modhera, C.D.; Desai, A.K. Effect of glass on strength of concrete subjected to sulphate attack. *Int. J. Civil Eng. Res. Dev.* **2011**, *1*, 1–13. Available online: https://papers.ssrn.com/sol3/papers.cfm?abstract_id=3501397 (accessed on 12 November 2020).
105. Nassar, R.; Soroushian, P. Field investigation of concrete incorporating milled waste glass. *J. Solid Waste Technol. Manag.* **2011**, *37*, 307–319. [CrossRef]
106. Shao, Y.; Lefort, T.; Moras, S.; Rodriguez, D. Studies on concrete containing ground waste glass. *Cem. Concr. Res.* **2000**, *30*, 91–100. [CrossRef]
107. Shayan, A.; Xu, A. Value-added utilisation of waste glass in concrete. *Cem. Concr. Res.* **2004**, *34*, 81–89. [CrossRef]
108. Shi, C.; Wu, Y.; Riefler, C.; Wang, H. Characteristics and pozzolanic reactivity of glass powders. *Cem. Concr. Res.* **2005**, *35*, 987–993. [CrossRef]
109. Kamali, M.; Ghahremaninezhad, A. Effect of glass powders on the mechanical and durability properties of cementitious materials. *Constr. Build. Mater.* **2015**, *98*, 407–416. [CrossRef]
110. Yamada, K.; Ishiyama, S. Maximum dosage of glass cullet as fine aggregate in mortar. In Proceedings of the International Conference on Achieving Sustainability in Construction, Dundee, UK, 5–6 July 2005; Dhir, R.K., Dyer, T.D., Newlands, M.D., Eds.; Thomas Telford: London, UK, 2005; pp. 185–192.
111. Jin, W.; Meyer, C.; Baxter, S. Glasscrete-concrete with glass aggregate. *ACI Mater. J.* **2000**, *97*, 208–213.
112. Idir, R.; Cyr, M.; Tagnit-Hamou, A. Use of fine glass as ASR inhibitor in glass aggregate mortars. *Constr. Build. Mater.* **2010**, *24*, 1309–1312. [CrossRef]
113. Xie, Z.; Xiang, W.; Xi, Y. ASR Potentials of Glass Aggregates in Water-Glass Activated Fly Ash and Portland Cement Mortars. *J. Mater. Civ. Eng.* **2003**, *15*. [CrossRef]
114. Singh, V.K. Effect of polypropylene fiber on properties of concrete. *Int. J. Eng. Sci. Res. Technol.* **2014**, *3*, 312–317.
115. Najimi, M.; Farahani, F.M.; Pourkhorshidi, A.R. Effects of polypropylene fibers on physical and mechanical properties of concretes. In Proceedings of the 3rd International Conference on Concrete & Development, CD03-009, Tehran, Iran, 27–29 April 2009; pp. 1073–1081. Available online: https://www.irbnet.de/daten/iconda/CIB13842.pdf (accessed on 20 February 2020).
116. Nili, M.; Afroughdaset, V. The effects The effects of silica fume and polypropylene fibers on the impact resistance and mechanical properties of concrete. *Constr. Build. Mater.* **2010**, *24*, 927–933. [CrossRef]
117. Satisha, N.S.; Shruthi, C.G.; Kiran, B.M.; Sanjith, J. Study on effect of polypropylene fibres on mechanical properties of normal strength concrete withpartial replacement of cement by GGBS. *Int. J. Adv. Res. Trends Eng. Technol.* **2018**, *5*, 23–51. Available online: www.ijartet.com (accessed on 10 January 2021).
118. Badogiannis, E.G.; Christidis, K.I.; Tzanetatos, G.E. Evaluation of the mechanical behavior of pumice lightweight concrete reinforced with steel and polypropylene fibers. *Constr. Build. Mater.* **2019**, *196*, 443–456. [CrossRef]
119. Szeląg, M. Evaluation of cracking patterns of cement paste containing polypropylene fibers. *Compos. Struct.* **2019**, *220*, 402–411. [CrossRef]
120. Afroughsabet, V.; Ozbakkaloglu, T. Mechanical and durability properties of high-strength concrete containing steel and polypropylene fibers. *Constr. Build. Mater.* **2015**, *94*, 73–82. [CrossRef]
121. Castoldi, R.S.; Souza, L.M.S.; Andrade Silva, F. Comparative study on the mechanical behavior and durability of polypropylene and sisal fiber reinforced concretes. *Constr. Build. Mater.* **2019**, *211*, 617–628. [CrossRef]
122. Alrshoudi, F.; Mohammadhosseini, H.; Md. Tahir, M.; Alyousef, R.; Alghamdi, H.; Alharbi, Y.R.; Alsaif, A. Sustainable Use of Waste Polypropylene Fibers and Palm Oil Fuel Ash in the Production of Novel Prepacked Aggregate Fiber-Reinforced Concrete. *Sustainability* **2020**, *12*, 4871. [CrossRef]
123. Rostami, R.; Zarrebini, M.; Mandegari, M.; Sanginabadi, K.; Mostofinejad, D.; Abtahi, S.M. The effect of concrete alkalinity on behavior of reinforcing polyester and polypropylene fibers with similar properties. *Cem. Concr. Compos.* **2019**, *97*, 118–124. [CrossRef]
124. Wadekar, A.P.; Pandit, R.D. Study of different types fibres used in high strength fibre reinforced concrete. *Int. J. Innov. Res. Adv. Eng.* **2014**, *1*, 225–230. Available online: www.ijirae.com (accessed on 10 January 2021).

125. Rangelov, M.; Nassiri, S.; Haselbach, L.; Englund, K. Using carbon fiber composites for reinforcing pervious concrete. *Constr. Build. Mater.* **2016**, *126*, 875–885. [CrossRef]
126. Afroughsabet, V.; Biolzi, L.; Ozbakkaloglu, T. High-performance fiber-reinforced concrete: A review. *J. Mater. Sci.* **2016**, *51*, 6517–6551. [CrossRef]
127. Zollo, R.F. Fiber-reinforced concrete: An overview after 30 years of development. *Cem. Concr. Compos.* **1997**, *19*, 107–122. [CrossRef]
128. Giurgiutiu, V. Chapter 5—Damage and Failure of Aerospace Composites. In *Structural Health Monitoring of Aerospace Composites*; Elsevier: Amsterdam, The Netherlands, 2016; pp. 125–175.
129. Kalifa, P.; Chene, G.; Galle, C. High-temperature behavior of HPC with polypropylene fibres. From spalling to microstructure. *Cem. Concr. Res.* **2001**, *31*, 1487–1499. [CrossRef]
130. Kosmatka, S.; Kerkhoff, B.; Panarese, W. *Design and Control of Concrete Mixtures*; PCA: Warsaw, Poland, 2009.
131. Naaman, A.E.; Reinhart, H.W. *High Performance Fiber Reinforced Cement Composites 2*; Taylor & Francis Group: Oxford, UK, 1996.

materials MDPI

Article

Study on the Technology of Monodisperse Droplets by a High-Throughput and Instant-Mixing Droplet Microfluidic System

Rui Xu, Shijiao Zhao, Lei Nie, Changsheng Deng, Shaochang Hao, Xingyu Zhao, Jianjun Li, Bing Liu and Jingtao Ma *

State Key Laboratory of New Ceramics and Fine Processing, Institute of Nuclear and New Energy Technology, Tsinghua University, Beijing 100084, China; xur17@mails.tsinghua.edu.cn (R.X.); zhaosj16@mails.tsinghua.edu.cn (S.Z.); niel18@mails.tsinghua.edu.cn (L.N.); changsheng@mail.tsinghua.edu.cn (C.D.); haosc@tsinghua.edu.cn (S.H.); zhaoxingyu@tsinghua.edu.cn (X.Z.); leejj@mail.tsinghua.edu.cn (J.L.); bingliu@mail.tsinghua.edu.cn (B.L.)
* Correspondence: majingtao@mail.tsinghua.edu.cn

Abstract: In this study, we report a novel high-throughput and instant-mixing droplet microfluidic system that can prepare uniformly mixed monodisperse droplets at a flow rate of mL/min designed for rapid mixing between multiple solutions and the preparation of micro-/nanoparticles. The system is composed of a magneton micromixer and a T-junction microfluidic device. The magneton micromixer rapidly mixes multiple solutions uniformly through the rotation of the magneton, and the mixed solution is sheared into monodisperse droplets by the silicone oil in the T-junction microfluidic device. The optimal conditions of the preparation of monodisperse droplets for the system have been found and factors affecting droplet size are analyzed for correlation; for example, the structure of the T-junction microfluidic device, the rotation speed of the magneton, etc. At the same time, through the uniformity of the color of the mixed solution, the mixing performance of the system is quantitatively evaluated. Compared with mainstream micromixers on the market, the system has the best mixing performance. Finally, we used the system to simulate the internal gelation broth preparation of zirconium broth and uranium broth. The results show that the system is expected to realize the preparation of ceramic microspheres at room temperature without cooling by the internal gelation process.

Keywords: microfluidic; high-throughput; micromixing; monodisperse droplets; internal gelation process

Citation: Xu, R.; Zhao, S.; Nie, L.; Deng, C.; Hao, S.; Zhao, X.; Li, J.; Liu, B.; Ma, J. Study on the Technology of Monodisperse Droplets by a High-Throughput and Instant-Mixing Droplet Microfluidic System. *Materials* **2021**, *14*, 1263. https://doi.org/10.3390/ma14051263

Academic Editors: Marcin Wachowski, Henryk Paul and Sebastian Mróz

Received: 14 February 2021
Accepted: 2 March 2021
Published: 7 March 2021

1. Introduction

Mixing is a necessary process for reactants to come into contact with each other before a reaction. A micromixer has the advantages of fast and uniform mixing, no contamination of the reagents and the reduction of reagent consumption [1,2], better heat and mass conduction, and can effectively realize chemical reactions sensitive to air and humidity [3] and the safer synthesis of dangerous compounds [4]. These advantages have attracted strong interest from researchers, leading to the widespread study of micromixers in DNA hybridization [5], cell activation [6,7], enzyme reactions [8], protein folding [9], water quality monitoring [10], flow chemistry [11] and the synthesis of micro-/nanoparticles, etc. [12–15].

Su, for example, in the synthesis of micro-/nanoparticles, used a T-junction mixer to mix two solutions to prepare 15–100 nm $BaSO_4$ nanoparticles [16]. Wang mixed the reagent and zirconium broth in a glass capillary, and silicone oil simultaneously sheared the two solutions to form droplets and prepared 100 μm ZrO_2 microspheres under the condition of zirconium broth flow rate of 1 μL/min [17]. Frenz embedded an external electrode on both sides of the microchannel to induce the fusion of two different component droplets through alternating current and prepared Fe_3O_4 nanoparticles smaller than 15 nm under the condition of an aqueous phase flow rate of 120 μL/h [18]. Zhang prepared a 1.8 mm wide and 100 mm thick micromagnetic gyromixer to achieve uniform mixing [19]. The micromixers adopted by the above researchers can be divided into passive micromixers and active micromixers according to whether there is an external power source. Passive micromixers mainly improve the mixing performance by increasing the contact area between fluids and constructing chaotic convection through the microchannels. Passive micromixers have a drawback, i.e., the mixing performance is not ideal when the Reynolds number is low, which limits their application. However, active micromixers do not have this drawback because they actively enhance mixing performance by using some form of external energy to generate chaotic convection, such as electric or magnetic fields. Active micromixers, for example micromagnetic gyromixers, also have limited application due to their complicated manufacturing process and high cost [20–22]. Moreover, by reducing the size of the droplets to shorten the distance of solute diffusion, researchers have currently achieved micromixing at a flow rate of μL/min or μL/h, which is unfavorable for large-scale industrial applications of micromixing. In addition, these micromixers usually have microchannels etched on polydimethylsiloxane and then thermally bonded together. If the two solutions react when mixed in the micromixer they may generate insoluble substances; for example, $FeCl_3$ solution and NaOH solution generate $Fe(OH)_3$ particles, which can easily block the microchannels and cause damage to the micromixer. However, a reusable, low-cost, high-throughput micromixer has not been developed yet. Therefore, there is an urgent need for a high-throughput micromixer with good mixing performance, which needs to be disassembled to clean the insoluble matter in the micromixer, so as to realize the reuse of the micromixer.

In this work, a novel high-throughput and instant-mixing droplet microfluidic system (noted as DMS) is constructed. The DMS is composed of an active magneton micromixer and a T-junction microfluidic device. Liquid droplets can be produced with such a device. The DMS can be used for the rapid and uniform mixing of two solutions, and can also be used for the preparation of micro/nano ceramic particles. It is easy to disassemble the DMS and clear the insoluble matter from the microchannel. By adjusting the structure of the T-junction microfluidic device and process parameters such as the magneton speed, and the content of surfactant, etc., the most suitable conditions for the DMS are found. The DMS is compared with the mainstream micromixers on the market and their mixing performance and the uniformity of the droplets' sizes are analyzed. The effectiveness of the DMS is evaluated by using dispersed phases of different viscosities to simulate the preparation of zirconium broth and uranium broth in the internal gelation process.

2. Materials and Methods

2.1. Construction of High-Throughput and Instant-Mixing Droplet Microfluidic System

The high-throughput and instant-mixing droplet microfluidic system is schematically shown in Figure 1. The DMS is mainly composed of a magneton micromixer and a coaxial T-junction (1/4-28UNF, Runze Fluid) microfluidic device. A charge coupled device camera (CCD, Olympus, Tokyo, Japan) is used to monitor the dropping processs in situ. Two miscible aqueous phases are propelled into the chamber equipped with magnets at a flow rate of mL/min through the injection path by two syringe pumps (XFP12-BD, Zhongxinqiheng, China). The mixing performance of the two aqueous phases is adjusted by controlling the magneton speed. In order to facilitate the flexible rotation of the magneton in the chamber, the shape of the chamber is set to a cylinder with an inner diameter of 8 mm and a height of 6 mm. The volume of the liquid filling the chamber can be estimated to be 0.256 mL by subtracting the volume of the magneton from the volume of the chamber. When the flow rate of the two aqueous phases is 0.25 mL/min and the flow time of the two aqueous phases in the microchannel is ignored, it takes only 30 s to for the fluid to fill the entire chamber. It means that it only takes 30 s for the two aqueous phases to mix thoroughly and form droplets. In addition, in order to prevent the magneton from sending the two aqueous phases without being sufficiently mixed into the sample path beforehand, this micromixer is designed with the injection path at the bottom of the chamber and the sample path at the top of the chamber. The mixed solution as the dispersed phase is sheared into droplets by the continuous phase of silicone oil in the coaxial T-junction microfluidic device and the droplets are collected in a measuring cylinder. The droplet formation process is observed under the CCD camera. The physical image of DMS and Magneton micromixer are mentioned in detail in Figure S1.

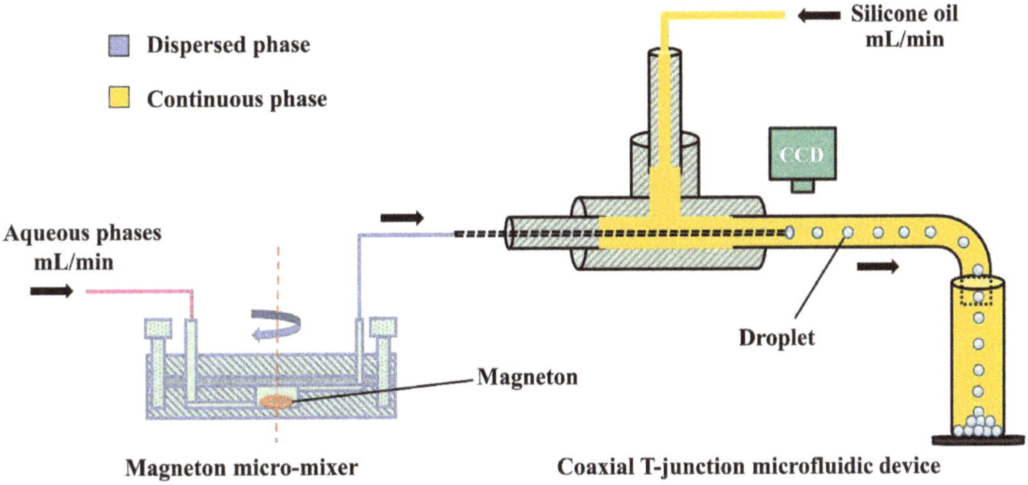

Figure 1. Schematic drawing of the droplet microfluidic system (DMS) for preparing monodisperse droplets.

The magneton micromixer is sealed with fastening screws and gaskets, and the pipe of the coaxial T-junction microfluidic device is fixed with inverted cone joints. When the two aqueous phases are not uniformly mixed or the mixing ratio is not appropriate to produce insoluble substances, the DMS can be easily disassembled to clean the clogged part, thereby realizing the reuse of the DMS and greatly reducing the cost compared with clogged and scrapped micromixer of the previous researchers.

The DMS increases the contact area of the two aqueous phases by using the magneton rotation to generate chaotic convection, and when the mixed solution is sheared into droplets, the solute diffusion distance is shortened to improve the mixing performance. Therefore, the DMS combines the characteristics of the active and passive micromixers.

2.2. Mixing-Target Liquids

In order to prevent the DMS from firstly clogging, deionized water was used as the aqueous phase in the process of preparing monodisperse droplets. In the comparison of mixing performance, two portions of 150 mL deionized water were added with 1 g of pigment. According to the mixing principle of the pigment, the same amount of the sky blue aqueous phase and the lemon yellow aqueous phase will become the dispersed phase of kelly green. The mixing performance is judged based on the color uniformity of the collected pictures by the CCD camera. In addition, a certain amount of polyvinyl alcohol abbreviated as PVA (Mw 13,000–23,000, Sigma-Aldrich (Munich, Germany)) was added to water with pigment to simulate a zirconium broth and a uranium broth. In order to prevent the addition of the pigment from affecting the formation of monodisperse droplets, the density, viscosity, and interfacial tension of different aqueous phases with the appropriate continuous phase are measured, as shown in Table 1. The composition of this suitable continuous phase, which will be given in the third part, is 83.6 mPa·s silicone oil (Aladdin, Shanghai, China) with 2% v/v Dow Corning 749 (Dow Corning, Midland, TX, USA).

Table 1. The density, viscosity, and interfacial tension of different aqueous phases.

Samples	Density (g/cm^3)	Viscosity (mPa·s)	Interfacial Tension (mN/m)
Deionized water	1.000	1.0	20.7
Water with lemon yellow pigment	1.013	8.1	20.5
Water with kelly green pigment	1.007	7.5	20.0
Water with sky blue pigment	1.001	7.8	20.2
Zirconium broth [23]	1.211	7.0	20.9
Uranium broth [24]	1.512	14.5	21.0
Water with PVA and kelly green pigment for simulating zirconium broth	1.003	8.1	19.6
Water with PVA and kelly green pigment for simulating uranium broth	1.008	15.2	20.0

It can be seen from Table 1 that the viscosity of the water and water with PVA after adding the pigment will increase a few mPa·s, and the density and interfacial tension are basically unchanged compared with the original solution. When the same amount of sky blue pigment solution is mixed with the lemon yellow pigment solution to obtain the kelly green pigment solution, compared with the two solutions before mixing, the density and interfacial tension of the kelly green pigment solution are basically unchanged, and the viscosity is slightly reduced. Zirconium broth and uranium broth are nearly saturated solutions, so it is normal that the density of water with PVA and pigment is lower than that of zirconium broth and uranium broth. Unlike macroflow, the interfacial tension plays a dominant role in microfluidics and the effect of gravity is usually negligible. In other words, the density difference between the simulated solution and the broth can be ignored in the DMS.

In general, deionized water is used as the aqueous phase and combined with the DMS to prepare monodisperse droplets. This mixing performance is quantitatively characterized by analyzing the color uniformity of the kelly green pigment solution mixed from the sky blue solution and the lemon yellow solution.

2.3. Characterization

The density of the aqueous phases is measured by a liquid densitometer. The viscosity of the aqueous phases is measured with an LVDV-1 digital rotation viscometer (Shanghai Jingtian Electronic Instrument Co., Ltd., Shanghai, China). The picture of the droplets formed at the capillary port and the picture of the droplets of the collecting cylinder are captured by an Olympus IX71 fluorescence microscope (Olympus, Tokyo, Japan). Through image recognition, the size and coefficient of variation of the droplets in these pictures are extracted, as shown in Figure 2A. The program automatically recognizes the number and outline of the droplets in the picture (Figure 2B), and calculates the size and coefficient of variation of the droplets. The specific identification principle is mentioned in the previous article by the research group [25]. The simplified variance normalization method is used to characterize the mixing performance (noted as *MP*) [1], as shown in Equation (1) where m_i is the gray value of the i-th point in the picture, and $\overline{m}$ is the average gray value on the picture, and n is the number of pixel points in the picture.

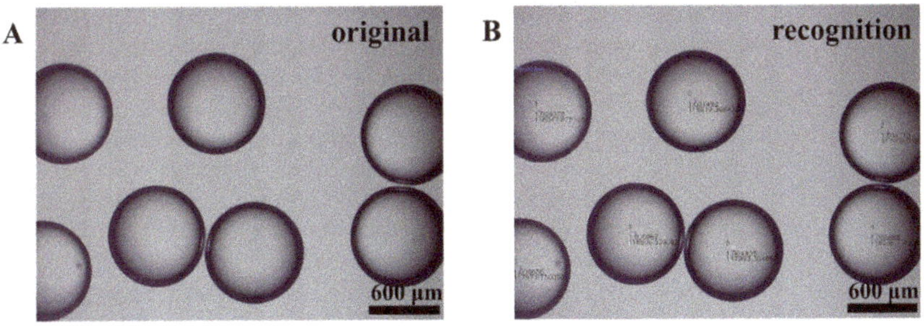

Figure 2. (**A**) The original image of droplets. (**B**) The image after program recognition.

The larger the value of *MP*, the better the mixing performance.

$$MP = 1 - \frac{1}{\overline{m}}\sqrt{\frac{\sum_{i=1}^{n}(m_i - \overline{m})^2}{n}} \quad (1)$$

3. Results and Discussion

3.1. The Structure of the T-Junction Microfluidic Device

The T-junction microfluidic device is a key device for forming monodisperse droplets. According to the flow direction of the continuous phase and the dispersed phase, the T-junction microfluidic device can be divided into T-junction perpendicular flow, T-junction transverse flow and coflowing. When the flow rate of dispersed phase and the continuous phase is 1 mL/min and 4 mL/min, respectively, and the continuous phase viscosity is 66 mPa·s, the droplets formed by the three flow structures are observed. It can be seen from Figure 3 that the size of the droplets prepared by the coaxial T-junction microfluidic device is the most uniform and the coefficient of variation is less than 5%, which meets the requirements of monodisperse droplets [26].

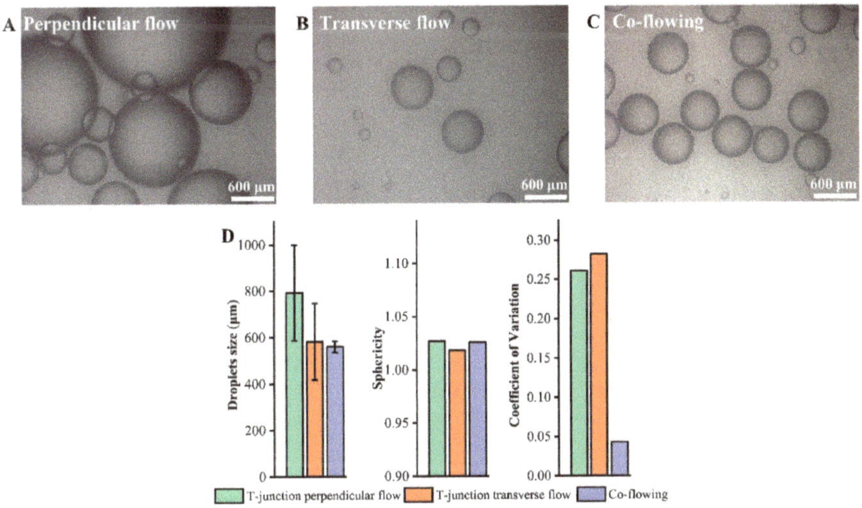

Figure 3. (**A**) T-junction perpendicular flow. (**B**) T-junction transverse flow. (**C**) Co-flowing. (**D**) Droplets' sizes and sphericity and coefficient of variation by the three structures.

The reason for the difference is related to the droplet formation mechanism of these three flow structures. T-junction perpendicular flow and T-junction transverse flow mainly use the pressure difference before and after the droplet to break the droplet. In T-junction perpendicular flow, the dispersed phase is squeezed into a continuous liquid column that moves in in the microchannel, resulting in uneven droplet sizes (Figure 3A). In T-junction transverse flow, the dispersed phase is squeezed into discontinuous liquid columns in the microchannel and satellite droplets are generated, resulting in uneven droplet sizes (Figure 3B). The mechanism of coflowing by the coaxial T-junction microfluidic device to generate droplets is Kelvin–Helmholtz instability. That is, when the difference between the flow rate of the two phases exceeds a certain range, monodisperse droplets will be generated, as shown in Figure 3C. In addition, due to the interfacial tension of the droplets and the continuous phase, the sphericity of the droplets formed by these three flow structures is all less than 1.05 (Figure 3D), which is good. The sphericity will not be discussed in the subsequent droplet evaluation process. Therefore, the suitable structure of the T-junction microfluidic device that generates monodisperse droplets is coflowing, that is the coaxial T-junction microfluidic device.

3.2. *The Magneton Rotation Speed*

The rotational speed of the magneton directly affects the mixing performance of the liquid in the chamber of the magneton micromixer. Equal amounts of water with lemon yellow pigment and water with sky blue pigment are pushed into the chamber to observe the mixing performance under different magneton speeds. The mixing performance at different magneton speeds can be seen from Figure 4. When the magneton speed is 0 r/min, the mixing performance reaches 0.800. This is solely due to the solute diffusion between the two pigment solutions. It can be seen from the picture of the chamber in the magnetic micromixer at 0 r/min that there are still some lemon yellow solutions that are not completely mixed. When the magneton speed increases from 0 r/min to 600 r/min, the mixing performance is significantly improved, reaching 0.944. However, when the magneton speed further increases from 600 r/min to 1200 r/min, the improvement of mixing performance reaches plateau. Therefore, 600 r/min was chosen as the magneton speed of this experiment.

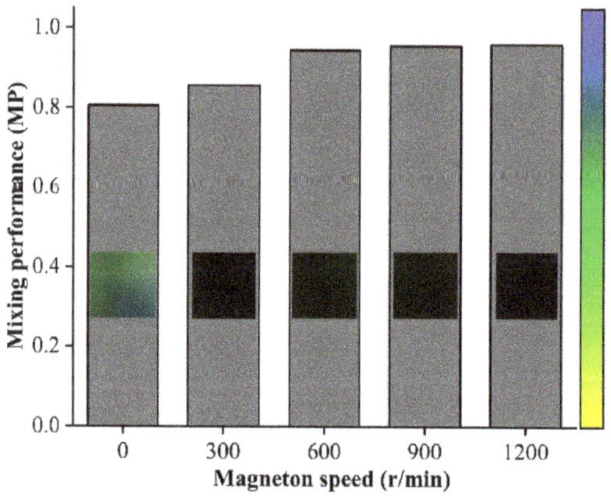

Figure 4. Mixing performance of different magneton speeds.

3.3. *The Content of Surfactant*

Dow Corning 749 (Dow Corning Co., Ltd.) is decamethyl-cyclopentasiloxane and trimethylated silica as the surfactant, which is one of the oil-soluble polymers. The surfactant is used to prevent coalescence between droplets by reducing the interfacial tension between the continuous phase and the dispersed phase. The surfactant is added to 50 mPa·s silicone oil as the continuous phase according to the volume ratio. When the flow rate of dispersed phase and the continuous phase is 0.5 mL/min and 2 mL/min, respectively, and the magneton speed is 600 r/min, the size and coalescence of droplets are observed, as shown in Figure 5. The droplets' sizes decrease with the increase in the content of surfactant. This can be attributed to the increase in the surfactant, which reduces the interfacial tension between the dispersed phase and the continuous phase. The decreased interfacial tension is beneficial to generation of smaller droplets by the DMS. The coefficient of variation of the droplets' sizes after the addition of surfactants does not have obvious regularities in Figure 5. It can be seen from the physical map of the collected droplets that the surfactant has little effect on the coalescence of the droplets. Then, in order to prevent the excessively high content of the surfactant from affecting the subsequent experiments of preparing monodisperse microspheres by DMS, 2% *v*/*v* of the surfactant was added to the silicone oil.

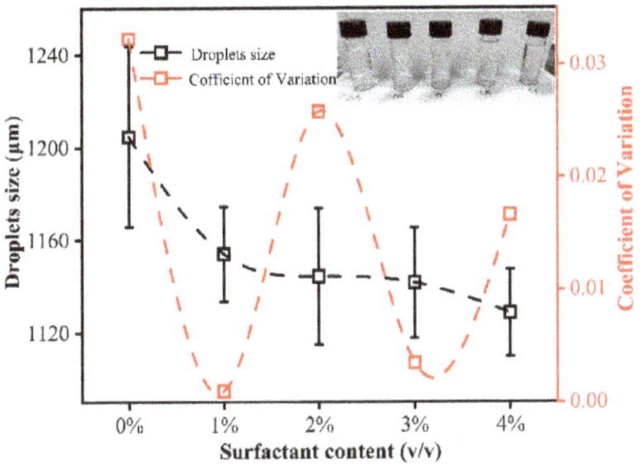

Figure 5. The variation of the content of surfactant with droplet's size and coefficient of variation.

3.4. Factors Affecting Droplets' Sizes and Coefficient of Variation

When 2% *v*/*v* of the surfactant is added to the silicone oil, the viscosity of the silicone oil needs to be re-measured as the viscosity of the continuous phase. When 2% *v*/*v* the surfactant is added to the silicone oil of different viscosity as the continuous phase, and the flow rate of the dispersed phase and the continuous phase is 0.5 mL/min and 2 mL/min, respectively, the magneton speed is 600 r/min, and the coaxial T-junction microfluidic device is used, the relationship between the droplet's size and coefficient of variation and various factors is obtained, as shown in Figure 6A. It can be seen that as the viscosity of continuous phase increases, the size of the droplets decreases, and the coefficient of variation decreases from the initial 0.047 to 0.003. This is because the viscosity of the continuous phase increases, which increases its shearing force on the dispersed phase, resulting in a smaller and more uniform droplet sizes. Then, when the viscosity of the continuous phase is very large, it will increase the flow resistance of the continuous phase and the dispersed phase in the microchannel, which is not conducive to high-throughput droplet microfluidic systems. Therefore, a moderate viscosity of continuous phase (83.6 mPa·s) was selected. From the physical image of the droplets collected, which formed at 83.6 mPa·s in a measuring cylinder (Figure 6A), it can be seen that the droplets are uniform in size and have good sphericity.

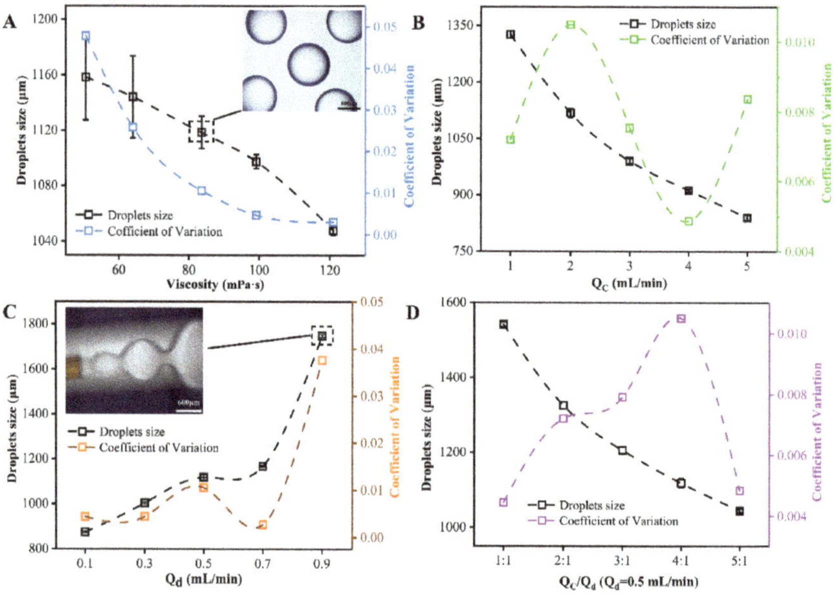

Figure 6. (**A**)The viscosity of continuous phase. (**B**) The flow rate of continuous phase (Q_c). (**C**) The flow rate of dispersed phase (Q_d). (**D**) The flow ratio of continuous phase to dispersed phase.

The effect of the flow rate of the continuous phase is investigated under the condition that the viscosity of the continuous phase is 83.6 mPa·s, the flow rate of the dispersed phase is 0.5 mL/min, 2% v/v surfactant is used, the magneton speed is 600 r/min, and the coaxial T-junction microfluidic device is used. The results are presented in Figure 6B. With the increase in the continuous phase flow rate, the droplet's size rapidly decreases and the coefficient of variation of the droplets fluctuates around 0.007. The volume of the continuous phase is only 100 mL on the syringe pump. Considering the initial bubble elimination time of the DMS and the time for the droplets to reach stable generation conditions, the DMS requires at least 50 min of running time to ensure accurate data collection. Therefore, 2 mL/min is chosen as the flow rate of continuous phase.

The influence of the flow rate of dispersed phase is investigated under the condition that the viscosity of the continuous phase is 83.6 mPa·s, the flow rate of the continuous phase is 2 mL/min, 2% v/v surfactant is used, the magneton speed is 600 r/min, and the coaxial T-junction microfluidic device is used. The results as shown in Figure 6C indicate that as the flow rate of the dispersed phase increases, the droplets' sizes also increase. When the flow rate of the dispersed phase is 0.9 mL/min, the droplets' size sin the microchannel become larger and the distance between the droplets becomes smaller, because the shear force of the continuous relative dispersed phase becomes smaller. This will easily cause the droplets of the microchannel to collide and the collected droplets in the measuring cylinder will be of uneven size. For example, when the flow rate of the dispersed phase is 0.9 mL/min, the coefficient of variation of the droplets will reach 0.04. However, if the flow rate of the droplets is small, it is not conducive to the high-throughput preparation of monodisperse droplets. Therefore, a moderate dispersed phase flow rate (0.5 mL/min) was selected.

The effect of the flow ratio of the continuous phase to the dispersed phase is analyzed under the condition that the viscosity of the continuous phase is 83.6 mPa·s, the flow rate of the dispersed phase is 0.5 mL/min, 2% v/v surfactant is used, the magneton speed is 600 r/min, and the coaxial T-junction microfluidic device is used, as shown in Figure 6D. As the flow ratio of the continuous phase to the dispersed phase becomes larger, the size of the droplets decreases rapidly, and the coefficient of variation of the droplets fluctuates at 0.01, which can be ignored. Since the inner diameter of the outlet tube of the microchannel is 1600 μm, the size of the droplets cannot be too large to avoid friction between the droplets and the tube wall of the microchannel. This means that the larger the flow ratio of the continuous phase to the dispersed phase, the better. However, considering the stable working time of the DMS, 4:1 was chosen as the best flow ratio of the continuous phase to the dispersed phase.

Therefore, using deionized water as the aqueous phase, the best conditions for the DMS to prepare monodisperse droplets are found—that is, in the coaxial T-junction microfluidic device, the magneton speed is 600 r/min, and the content of surfactant is 2% v/v, the viscosity of continuous phase is 83.6 mPa·s, the flow rate of continuous phase and dispersed phase is 2 mL/min and 0.5 mL/min, respectively, and the flow ratio of the continuous phase to the dispersed phase is 4:1.

3.5. Correlation Analysis of Influencing Factors

As can be seen from the above, there are many factors that affect the size of the droplets prepared by the DMS. Mathematical statistics and numerical simulations are widely used in scientific research [27–29]. The SPSS software (version 22.0) developed by IBM (Armonk, NY, USA) is used to do Pearson correlation analysis of these influencing factors and the droplets' sizes. The correlation coefficient and significance of each influencing factor are shown in Table 2. It can be seen that except for the flow rate of the dispersed phase, the other influencing factors are negatively related to the droplets' sizes. The flow ratio of the continuous phase to the dispersed phase is varied by maintaining the flow rate of the dispersed phase at a constant rate while the flow rate of the continuous phase is changed. Its correlation coefficient is basically the same as that of the flow rate of the continuous phase; both are about 0.974. In addition, the viscosity of the continuous phase has the largest correlation coefficient, followed by the flow rate of the continuous phase, then the flow rate of the dispersed phase and finally the content of the surfactant. This shows that the viscosity of the continuous phase has the closest relationship with the droplets' sizes in the DMS, which is significant at the level of 0.01. Therefore, the droplets' sizes can be changed mainly by changing the viscosity and flow rate of the continuous phase.

Table 2. Pearson correlation coefficient of influencing factors and droplet size.

The Influencing Factors	Pearson Correlation with Droplets' Size
The content of surfactant	-0.886^{α}
The viscosity of continuous phase	-0.987^{β}
The flow rate of continuous phase	-0.973^{β}
The flow rate of dispersed phase	0.900^{α}
The flow ratio of continuous phase to dispersed phase	-0.974^{β}

$^{\alpha}$ Correlation is significant at the 0.05 level (2-tailed); $^{\beta}$ Correlation is significant at the 0.01 level (2-tailed).

3.6. Mixing Performance and Uniformity of Droplets' Sizes

In order to verify the mixing performance of the DMS, the DMS is compared with the mainstream micromixers on the market. The aqueous solutions of sky blue pigment and lemon yellow pigment (as shown in Table 1) are respectively passed into the micromixer at the dispersed phase and are sheared into droplets under the abovementioned optimal droplet generation conditions. The droplet generation drawings in the microchannel drawing are shown in Figure 7.

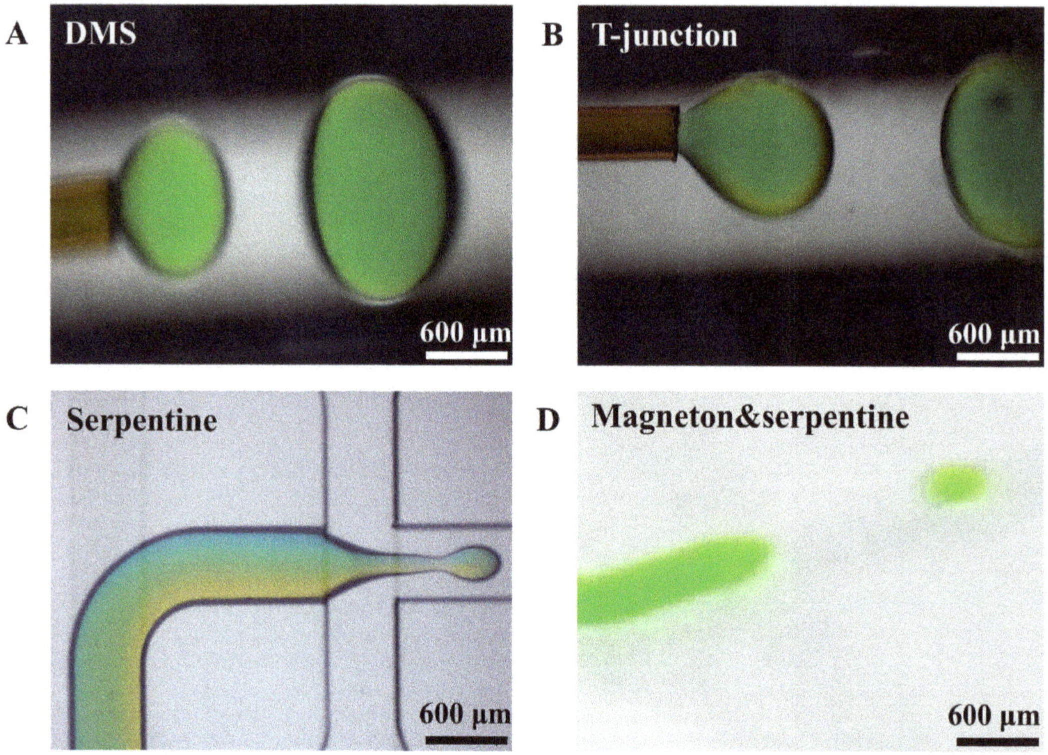

Figure 7. (**A**) DMS. (**B**) T-junction microfluidic device. (**C**) Serpentine micromixing chip. (**D**) Magneton and serpentine microfluidic device.

Figure 7A shows that the droplets prepared by DMS are evenly mixed in color, and Figure 7B shows that the droplets prepared by the T-junction microfluidic device, which is composed of a T-junction micromixer and the coaxial T-junction microfluidic device, are yellow on the outermost surface and have uniform internal color. The colors of the two pigment solutions in the droplets prepared by the serpentine micromixing chip in Figure 7C are still clearly visible. Moreover, the two pigment solutions have gone through eight u-shaped bends in the serpentine micromixing chip and they are still not evenly mixed. Their mixing performance is quantitatively extracted, as shown in Figure 8A. It can be seen from Figure 8A that the mixing performance of the collected droplets in a measuring cylinder is better than that of the generated droplets in the microchannel because the solute diffusion inside the droplets increases mixing performance when the generated droplets move into the measuring cylinder through the microchannel. Under the same flow rate of the dispersed phase, whether it is the generated droplets or the collected droplets, the mixing performance of DMS is the best, followed by the T-junction microfluidic device, and the serpentine micromixing chip comes last. This is because both T-junction microfluidic device and serpentine micromixing chip are passive micromixers, which mainly rely on solute molecular diffusion for mixing, while DMS is an active micromixer, which is prone to generating chaotic convection and has the best mixing performance among these microfluidic devices.

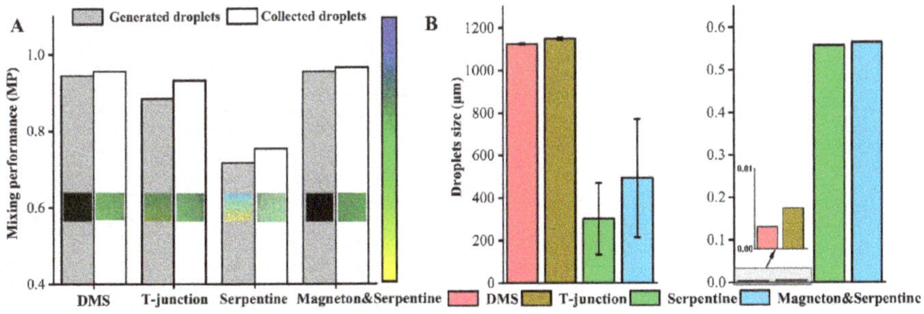

Figure 8. (**A**) Mixing performance of different microfluidic devices. (**B**) Droplets size and coefficient of variation of different microfluidic devices.

The magneton and serpentine microfluidic device in Figure 7D is integrated by the magneton micromixer and a serpentine micromixing chip. When the magneton speed is 600 r/min, the magneton and serpentine microfluidic device and the DMS are considered to reach the same degree of mixing. This is explained from the fact that the mixing performance of the two is similar in Figure 8A. However, the magneton and serpentine microfluidic device cannot achieve high throughput. Under the conditions of 0.5 mL/min for the dispersed phase and 2 mL/min for the continuous phase, the dispersed phase cannot be sheared into droplets by the continuous phase. The continuous phase can flow countercurrently into the chamber with the magneton that mixes the aqueous phase solutions, hindering the mixing of the solutions, resulting in the inability to form droplets. Only when the flow rate of the dispersed phase is 20 μL/min and the flow rate of the continuous phase is 100 μL/min can the droplets be formed. At a similar mixing performance level, the DMS can achieve a flow rate of mL/min to prepare droplets while the magneton and serpentine microfluidic device can only prepare droplets at μL/min, which directly illustrates the superiority of the DMS structure. It can be seen from Figure 8B that the size of the droplets generated by the DMS and the T-junction microfluidic device is very uniform, and the coefficient of variation is below 0.006. The coefficient of variation of the droplets by the serpentine micromixing chip and magneton and serpentine microfluidic device exceeded 0.5, indicating that the size distribution of the droplets prepared by these two micromixers is very large.

Therefore, compared to these three kinds of micromixers, the DMS has the best mixing performance and droplet size uniformity. In principle, the coaxial T-junction microfluidic device prepares droplets which are more stable under force on a three-dimensional flow scale than other microfluidic devices. The magneton micromixer has a simple structure and uniform mixing. It makes the flow resistance in the DMS far smaller than that of the complex microchannels in other microfluidic devices, thereby achieving the preparation of high-throughput uniformly mixed monodisperse droplets.

3.7. Simulated Broths Experiment by DMS

In the preparation of ZrO_2 and UO_2 gel microspheres by the internal gelation process, the metal ion solution needs to be mixed with a mixed solution of urea and hexamethylenetetramine (noted as HMUR solution) to form a zirconium or uranium broth, and the broth is rapidly dispersed into droplets and falls into hot silicone oil to form gel microspheres [30,31], which can be found in Figure S2. However, hexamethylenetetramine is thermally unstable and easily decomposes into ammonium hydroxide, leading to the premature precipitation and gelation of metal ions to block the dispersing device. The reaction can be simplified into Equation (2) [17,30,31] and M^{n+} is the metal ion. Therefore, the metal ion solution and the HMUR solution need to be cooled and mixed at 5 °C, which is not conducive to large-scale industrial production. Moreover, when the metal ion solution and the HMUR solution are not sufficiently mixed, certain positions in the mixed solution will reach the pH at which the metal ion gelation reaction occurs, so the broth will quickly gel and block the device.

$$M^{n+} + nOH^{-} \rightarrow M(OH)_n \quad (2)$$

The water with PVA and sky blue pigment solution is used to simulate metal ion solution and the water with PVA and lemon yellow pigment solution is used to simulate HMUR solution. The two kinds of solutions are mixed into the kelly green pigment solution similar to the interfacial tension and viscosity of the broth by the magneton micromixer, and then droplets are formed on the coaxial T-junction microfluidic device under the abovementioned optimal droplet generation conditions. The pictures of the collected droplets in the measuring cylinder are shown in Figure 9A,B. The droplet size and the coefficient of variation and the mixing performance of the DMS for these two simulated broths can be seen from Figure 9C.

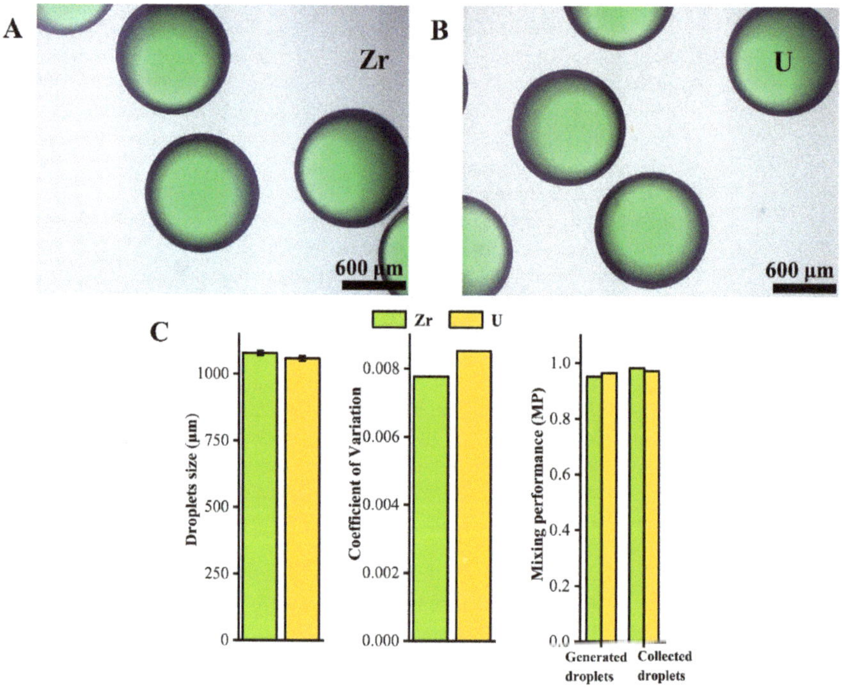

Figure 9. (**A**) Collected droplets of simulated zirconium broth by DMS. (**B**) Collected droplets of simulated uranium broth by DMS. (**C**) Characterization of droplets.

It can be seen from the Figure 9A,B that the droplets of simulated zirconium broth and simulated uranium broth are formed by DMS with good sphericity and uniform size and uniform mixing. The coefficient of variation of the droplets is less than 0.01. The mixing performance is greater than 0.95 for the generated droplets in the microchannel and the collected droplets in the measuring cylinder, which can be regarded as uniformly mixed. In addition, the size of the droplets formed by the simulated uranium broth is slightly smaller than that of the simulated zirconium broth. The viscosity of the simulated uranium broth is greater than that of the simulated zirconium broth. However, previous studies have shown that changing the viscosity of the dispersed phase will not change the force state of the droplets in the microchannel. So, it is normal that the size of the droplets slightly changes with the viscosity of the dispersed phase [32].

The internal gelation process is a process of hydrolysis of metal ions which is heavily dependent on temperature. When the temperature rises from 5 °C to 20 °C, the protonation of HMTA and the decomposition of protonated HMTA will accelerate, leading to an increase in the pH of the broth and promoting the hydrolysis of the metal ions [23,24]. For example, when the temperature rises from 5 °C to 20 °C, the stability time of the zirconium broth is reduced from 5 h to 1 h, and the stability time of uranium broth is reduced from 16 h to 200 s [23,24]. Because of the short stabilization time, it is difficult to achieve a continuous internal gelation process at room temperature without cooling to prepare the zirconia or uranium oxide microspheres. The time for DMS to mix and form droplets is only 30 s, much less than 200 s and 1 h. Therefore, it is expected that the mixing process using the DMS will not strongly affect the hydrolysis and gelation process of the metal ions such that gel microspheres can be prepared at room temperature by DMS without cooling the precursor solution.

4. Conclusions

In this study, a novel high-throughput and instant-mixing droplet microfluidic system is designed for solution mixing and preparation of micro-/nanoparticles. The system is detachable and it is easy to clean any blockages in the microchannel, which realizes the reuse of the system and greatly reduces the cost. Moreover, the system can mix the solution uniformly and produce droplets of uniform size at a flow rate of mL/min, which overcomes the shortcomings of low droplet yield and easy clogging of the micromixing chips on the market.

The results show that the best conditions for the DMS to prepare uniform mixing and monodisperse droplets with good sphericity are in the coaxial T-junction microfluidic device, where the magneton speed is 600 r/min, and the content of surfactant is 2% v/v, the viscosity of continuous phase is 83.6 mPa·s, the flow rate of continuous phase and dispersed phase is 2 mL/min and 0.5 mL/min, respectively, and the flow ratio of continuous phase to dispersed phase is 4:1. The viscosity and flow rate of continuous phase have a major impact on monodisperse droplets of different sizes. The DMS achieves the preparation of monodisperse droplets with better mixing performance than three micromixing chips on the market. Moreover, the simulation broths are used to simulate the preparation of zirconium and uranium gel microspheres in the internal gelation process by the DMS. The DMS can potentially realize the continuous production of ZrO_2 and UO_2 ceramic microspheres without cooling at room temperature. Thus, the DMS is expected to meet the demands in various fields, including the high-volume industrialization of microfluidics, micromixing, and micro-/nanoparticles.

Supplementary Materials: The following are available online at https://www.mdpi.com/1996-1944/14/5/1263/s1, Figure S1: (A) Physical image of DMS; (B) The physical image of the magneton micromixer. Figure S2: (A) The external gelation process; (B) The internal gelation process.

Author Contributions: Conceptualization, R.X., S.H. and J.M.; methodology, R.X. and J.M.; software, R.X.; validation, R.X.; formal analysis, R.X.; investigation, R.X.; resources, S.Z., L.N., S.H., X.Z., J.L. and J.M.; data curation, R.X.; writing—original draft preparation, R.X.; writing—review and editing, C.D. and J.M.; visualization, R.X.; supervision, J.M. and C.D.; project administration, J.M.; funding acquisition, J.M. and B.L.; All authors have read and agreed to the published version of the manuscript.

Funding: This work was supported by National Natural Science Foundation of China (No.52073157) and "The Thirteenth Five-Year Plan" Discipline Construction Foundation of Tsinghua University (No.2017HYYXKJS1), and the National S&T Major Project (No.ZX06901-25).

Institutional Review Board Statement: The study was conducted according to the guidelines of the Declaration of Helsinki, and approved by the Institutional Review Board of Institute of Nuclear and New Energy Technology, Tsinghua University. (not appliable).

Informed Consent Statement: Informed consent was obtained from all subjects in-volved in the study.

Data Availability Statement: Data is contained within the article or supplementary material. The data presented in this study are available in [Study on the Technology of Monodisperse Droplets by a High-Throughput and Instant-Mixing Droplet Microfluidic System].

Conflicts of Interest: The authors declare no conflict of interest.

References

1. Park, S.J.; La, M.; Cha, K.J.; Kim, D.S. Development of contaminant-free and effective micro-mixing methods based on non-contact dispensing system. *Microelectron. Eng.* **2013**, *111*, 175–179. [CrossRef]
2. Chen, X.; Liu, S.; Chen, Y.; Wang, S. A review on species mixing in droplets using passive and active micromixers. *Int. J. Environ. Anal. Chem.* **2019**, *101*, 422–432. [CrossRef]
3. Joanicot, M.; Ajdari, A. Droplet Control for Microfluidics. *Science* **2005**, *309*, 887–888. [CrossRef]
4. Hartman, R.L.; Jensen, K.F. Microchemical systems for continuous-flow synthesis. *Lab Chip* **2009**, *9*, 2495–2507. [CrossRef]
5. Ufer, A.; Sudhoff, D.; Mescher, A.; Agar, D.W. Suspension catalysis in a liquid–liquid capillary microreactor. *Chem. Eng. J.* **2011**, *167*, 468–474. [CrossRef]
6. Patnaik, P.R. Microbioreactors for Cell Cultures: Analysis, Modeling, Control, Applications and Beyond. *Int. J. Bioautomation* **2015**, *19* (Suppl. 1), S1–S42.
7. Toda, K.; Ebisu, Y.; Hirota, K.; Ohira, S.I. Membrane-based microchannel device for continuous quantitative extraction of dissolved free sulfide from water and from oil. *Anal. Chim. Acta* **2012**, *741*, 38–46. [CrossRef]
8. Serra, C.A.; Chang, Z. Microfluidic-assisted synthesis of polymer particles. *Chem. Eng. Technol.* **2008**, *31*, 1099–1115. [CrossRef]
9. Dendukuri, D.; Doyle, P.S. The Synthesis and Assembly of Polymeric Microparticles Using Microfluidics. *Adv. Mater.* **2010**, *21*, 4071–4086. [CrossRef]
10. Elmas, S.; Pospisilova, A.; Sekulska, A.A.; Vasilev, V.; Nann, T.; Thornton, S.; Priest, C. Photometric Sensing of Active Chlorine, Total Chlorine, and pH on a Microfluidic Chip for Online Swimming Pool Monitoring. *Sensors* **2020**, *20*, 3099. [CrossRef]
11. Reis, M.H.; Leibfarth, F.A.; Pitet, L.M. Polymerizations in Continuous Flow: Recent Advances in the Synthesis of Diverse Polymeric Materials. *ACS Macro Lett.* **2020**, *9*, 123–133. [CrossRef]
12. Bhatia, S.N.; Ingber, D.E. Microfluidic organs-on-chips. *Nat. Biotechnol.* **2014**, *32*, 760–772. [CrossRef] [PubMed]
13. Koester, S.; Angile, F.E.; Duan, H.; Agresti, J.J.; Wintner, A.; Schmitz, C.; Rowat, A.C.; Merten, C.A.; Pisignano, D.; Griffiths, A.D.; et al. Drop-based microfluidic devices for encapsulation of single cells. *Lab Chip* **2008**, *8*, 1110–1115. [CrossRef]
14. Yi, C.; Li, C.W.; Ji, S.; Yang, M. Microfluidics technology for manipulation and analysis of biological cells. *Anal. Chim. Acta* **2006**, *560*, 1–23. [CrossRef]
15. Chen, Y.; Wang, Y.J.; Yang, L.M.; Luo, G.S. Micrometer-sized monodispersed silica spheres with advanced adsorption properties. *AIChE J.* **2010**, *54*, 298–309. [CrossRef]
16. Su, Y.F.; Kim, H.; Kovenklioglu, S.; Lee, W.Y. Continuous nanoparticle production by microfluidic-based emulsion, mixing and crystallization. *J. Solid State Chem.* **2007**, *180*, 2625–2629. [CrossRef]
17. Wang, P.; Jiang, L.; Nunes, J.; Hao, S.; Chen, H. Droplet Micro-Reactor for Internal Gelation to Fabricate ZrO2 Ceramic Microspheres. *J. Am. Ceram. Soc.* **2017**, *100*, 41–48. [CrossRef]
18. Frenz, L.; El Harrak, A.; Pauly, M.; Bégin-Colin, S.; Griffiths, A.D.; Baret, J.C. Droplet-based microreactors for the synthesis of magnetic iron oxide nanoparticles. *Angew. Chem.* **2010**, *47*, 6817–6820. [CrossRef] [PubMed]
19. Zhang, Y.; Wang, T.-H. Micro magnetic gyromixer for speeding up reactions in droplets. *Microfluid. Nanofluid.* **2012**, *12*, 787–794. [CrossRef] [PubMed]
20. Ward, K.; Fan, Z.H. Mixing in microfluidic devices and enhancement methods. *J. Micromech. Microeng.* **2015**, *25*, 094001. [CrossRef]
21. Lee, C.Y.; Chang, C.L.; Wang, Y.N.; Fu, L.M. Microfluidic Mixing: A Review. *Int. J. Mol. Sci.* **2011**, *12*, 3263–3287. [CrossRef]
22. Hessel, V.; Lowe, H.; Schonfeld, F. Micromixers—A review on passive and active mixing principles. *Chem. Eng. Sci.* **2005**, *60*, 2479–2501. [CrossRef]
23. Gao, Y.; Ma, J.; Zhao, X.; Hao, S.; Deng, C.; Liu, B.; Franks, G. An Improved Internal Gelation Process for Preparing ZrO2 Ceramic Microspheres without Cooling the Precursor Solution. *J. Am. Ceram. Soc.* **2015**, *98*, 2732–2737. [CrossRef]
24. Li, S.; Bai, J.; Cao, S.; Yin, X.; Tan, C.; Li, P.; Tian, W.; Wang, J.; Guo, H.; Qin, Z. An Improved Internal Gelation Process without Cooling the Solution for Preparing Uranium Dioxide Ceramic Microspheres. *Ceram. Int.* **2018**, *44*, 2524–2528. [CrossRef]
25. Xu, R.; Chen, J.; Zhao, S.; Hao, S.; Zhao, X.; Li, J.; Deng, C.; Ma, J. Preparation of monodisperse ZrO2 ceramic microspheres (>200 μm) by coaxial capillary microfluidic device assisted internal gelation process. *Ceram. Int.* **2019**, *45*, 19627–19634.
26. Lewis, P.C.; Graham, R.R.; Nie, Z.; Xu, S.; Seo, M.; Kumacheva, E. Continuous Synthesis of Copolymer Particles in Microfluidic Reactors. *Macromolecules* **2005**, *38*, 4536–4538. [CrossRef]
27. Abbasnia, S.; Nasri, Z.; Najafi, M. Comparison of the mass transfer and efficiency of Nye tray and sieve tray by computational fluid dynamics. *Sep. Purif. Technol.* **2019**, *215*, 276–286. [CrossRef]
28. Valipour, P.; Ghasemi, S.E.; Khosravani, M.R.; Ganji, D.D. Theoretical analysis on nonlinear vibration of fluid flow in single-walled carbon nanotube. *J. Theor. Appl. Phys.* **2016**, *10*, 211–218. [CrossRef]
29. Bagheri, M.; Azmoodeh, M. Substrate Stiffness Changes Cell Rolling and Adhesion over L-selectin Coated Surface in a Viscous Shear Flow. *arXiv* **2019**, arXiv:1910.00002.
30. Sood, D.D. The role sol–gel process for nuclear fuels-an overview. *J. Sol Gel Sci. Technol.* **2011**, *59*, 404–416. [CrossRef]

31. Arima, T.; Idemitsu, K.; Yamahira, K.; Torikai, S.; Inagaki, Y. Application of internal gelation to sol-gel synthesis of ceria-doped zirconia microspheres as nuclear fuel analogous materials. *J. Alloys Compd.* **2005**, *394*, 271–276. [CrossRef]
32. Cramer, C.; Fischer, P.; Windhab, E.J. Drop formation in a co-flowing ambient fluid. *Chem. Eng. Sci.* **2004**, *59*, 3045–3058. [CrossRef]

materials

MDPI

Article

A New Method for Modeling the Cyclic Structure of the Surface Microrelief of Titanium Alloy Ti6Al4V After Processing with Femtosecond Pulses

Volodymyr Hutsaylyuk [1,*], Iaroslav Lytvynenko [2], Pavlo Maruschak [2], Volodymyr Dzyura [2], Georg Schnell [3] and Hermann Seitz [3]

1 Institute of Robots and Machine Design, Military University of Technology, Gen. S. Kaliskiego str. 2, 00-908 Warsaw, Poland

2 Department of Industrial Automation, Ternopil National Ivan Puluj Technical University, Ruska str. 56, 46001 Ternopil, Ukraine; d_e_l@i.ua (I.L.); maruschak.tu.edu@gmail.com (P.M.); volodymyrdzyura@gmail.com (V.D.)

3 Microfluidics, Faculty of Mechanical Engineering and Marine Technology, University of Rostock, Justus-von-Liebig-Weg 6, 18059 Rostock, Germany; georg.schnell@uni-rostock.de (G.S.); hermann.seitz@uni-rostock.de (H.S.)

* Correspondence: volodymyr.hutsaylyuk@wat.edu.pl; Tel.: +48-22-261-839-245

Received: 5 October 2020; Accepted: 3 November 2020; Published: 5 November 2020

check for updates

Abstract: A method of computer modeling of a surface relief is proposed, and its efficiency and high accuracy are proven. The method is based on the mathematical model of surface microrelief, using titanium alloy Ti6Al4V subjected to processing with femtosecond pulses as an example. When modeling the examples of microrelief, changes in the shape of segments-cycles of the studied surface processes, which correspond to separate morphological formations, were taken into account. The proposed algorithms were realized in the form of a computer simulation program, which provides for a more accurate description of the geometry of the microrelief segments. It was proven that the new method significantly increases the efficiency of the analysis procedure and processing of signals that characterize self-organized relief formations.

Keywords: titanium alloy Ti6Al4V; implant microrelief; mathematical model; profilometry signals

1. Introduction

It is known that during laser treatment of metals used for biomedical purposes (manufacture of implants), complex physicochemical processes take place in the surface layer, the kinetics of which is determined by the amount of energy introduced into the material and the processing time [1]. Periodic titanium patterns induce more uniform and direct cell growth. This effect is mainly connected with the surface properties of textured titanium implants [2,3]. When selected correctly, these parameters make it possible to form a surface layer of metals with a predetermined structure, grain size, phase composition, hardness and surface roughness [4]. However, estimating the parameters of a processed relief based on roughness alone is an approximate method that allows detecting the integral value of roughness only [5–7].

In this case, the geometric features of the relief formed, even with the same roughness values, may differ due to local peculiarities of micro- and macro-stresses distributed in the area of laser processing of materials [8,9]. Currently, investigations are ongoing into femtosecond laser treatment of titanium alloys, which is superior to the previously studied nanosecond laser treatment, as it makes it possible to create a surface topography of different geometries. It is known that increased overlap of laser-treated sections leads to an increased periodicity (cyclicity) of the treated surface microrelief. Femtosecond treatment of the material provides for the formation of highly ordered conical microstructures on the surface [10].

Modern methods of profilometry allow conducting functional diagnostics of the surface condition and detecting ordered structural formations. However, it remains important to create effective computer diagnostic systems for the automated processing of the received signals, followed by a preliminary diagnostic conclusion about the surface condition. Thus, the development of approaches for modeling the surface microgeometry of titanium alloys is the basis for creating a technology for making a self-organized microrelief with an optimal rate of osteointegration and modifying its surface with high-energy pulses [11–13].

Improving the methods of modeling the microgeometry of a laser-treated surface is one of the steps for ensuring the reproducibility of the relief formed and creating methods for describing its diagnostic state. Both parametric (roughness criteria) and nonparametric evaluation criteria are used to evaluate microgeometry. In particular, some works are known, the authors of which use densities and density functions describing the distribution of ordinates and tangents of the profile inclination angles, as well as profile microtopography. The effectiveness of the nonparametric approach to solving the optimization problems has been proven by numerous studies. In particular, earlier in [4], a mathematical model of the cyclic structure of the surface microrelief of the titanium alloy Ti6Al4V was described. The self-organization of the surface subjected to the impact-oscillatory laser effect was considered to be a cyclic random process, which provides for a description of the geometric features of the microrelief [14,15]. The components of the proposed model take into account the segments-cycles of cyclic microrelief.

The purpose of this research is to develop a new mathematical model that allows taking into account the amplitude parameters at each segment-cycle within the microrelief structure of the surface of titanium alloy Ti6Al4V and compare the results of the microrelief modeling using the new and known mathematical models.

2. Materials and Methods

A model for creating relief formations on the surface of titanium alloy Ti6Al4V, which were polished to a roughness s = 0.065 ± 0.003 µm and treated with a laser, was considered in this article [4,10]. A Yb-doped fiber laser system of the type UFFL_60_200_1030_SHG from Active Fiber Systems GmbH, Jena, Germany, with a pulse duration of 300 fs was used in this research. The system enabled a pulse repetition rate from 50.3 kHz up to 18.6 MHz with a maximum pulse energy of 200 µJ. The linear polarized Gaussian beam was deflected by an intelliSCAN 14 scan head (Scanlab GmbH, Puchheim, Germany) and focused with an f-theta lens with a focal length of 163 mm, resulting in a circular focus diameter of d_f = 36 µm (1/e^2). For the experimental work, a wavelength of 1030 nm was applied. Laser structuring was carried out on a Microgantry GU4 micromachining center from Kugler GmbH, Salem, Germany. This spot diameter was used for all laser parameter calculations [4,10]. The findings presented in article [4], in which pulse treatment was performed by overlapping the areas of laser treatment (LO), were the experimental basis of this research. High values of the parameters considered caused phase transformations and heat accumulation, which led to a reduced ablation threshold and increased roughness. In this paper, the surface relief properties of titanium alloy Ti6Al4V were modeled by LO 40% at a laser fluence of q = 4.91 J/cm^2, Figure 1, to verify the method efficiency. Both trenches and ridges were strongly covered with melt and nanoprotrusions due to a high level of fluence far above the ablation threshold of the material, resulting in a pronounced heat accumulation [4,10].

(a) (b)

Figure 1. SEM images of structured Ti6Al4V surface with a laser pulse overlap of 50% and scanning line overlap of 40% at a fluence of q = 4.91 J/cm^2. (**a**) A clear formation of trenches and ridges can be observed as a result of a low scanning line overlap (**b**) [4,10].

A generalized theoretical and methodological approach, which consisted of identifying the segmental structure of cyclic signals with a variable rhythm, was applied to the analysis of the surface relief, allowing us to process experimental data as part of the stochastic approach.

3. A New Mathematical Model of Cyclic Microrelief

The profilogram of the surface microrelief of the titanium alloy Ti6Al4V after femtosecond pulse treatment was considered as a stochastic cyclic process. In [14], the definition of a cyclic random process is given. It is a separable random process $\xi(\omega, l), \omega \in \mathbf{\Omega}, l \in [0, L)$, which is called a cyclic random process of a continuous argument, provided that there is the function $T(l, n)$, which satisfies the conditions of the rhythm function. In addition, finite-dimensional vectors $(\xi(\omega, l_1), \xi(\omega, l_2), \ldots, \xi(\omega, l_k))$ and $(\xi(\omega, l_1 + T(l_1, n)), \xi(\omega, l_2 + T(l_2, n))), \ldots, \xi(\omega, l_k + T(l_k, n)), n \in \mathbf{Z}$, where $\{l_1, \ l_2, \ldots, l_k\}$ is the set of the process separability, $\xi(\omega, l), \omega \in \mathbf{\Omega}, l \in [0, L)$ are stochastically equivalent in a broad sense for all integers $k \in \mathbf{N}$ where $T(l, n)$ is the rhythm function of the cyclic process, which reflects the regularities in the variation of temporal (spatial, in our case) intervals between its single-phase values. The main properties of this function are described in [15].

Figure 2 presents a block diagram of the well-known approach to modeling the surface microrelief [14]. Accordingly, the mathematical model of the cyclic surface microrelief was presented in a form that takes into account its segmental cyclic structure:

$$\xi_\omega(l) = \sum_{i=1}^{C} f_i(l), \ l \in \mathbf{W} \tag{1}$$

where C is the number of segments-cycles of the cyclic microrelief, $\mathbf{W}$ is the definition area of the cyclic microrelief, while the region of its values in case of the stochastic approach is the Hilbert space of random variables given on a single probabilistic space $(\xi_\omega(l) \in \mathbf{\Psi} = \mathbf{L}_2(\mathbf{\Omega}, \mathbf{P}))$. In expression (1), segments-cycles of the cyclic microrelief process are determined by indicator functions, that is:

$$f_i(l) = \xi_\omega(l) \times I_{\mathbf{W}_i}(l), \ i = \overline{1, C}, \ l \in \mathbf{W} \tag{2}$$

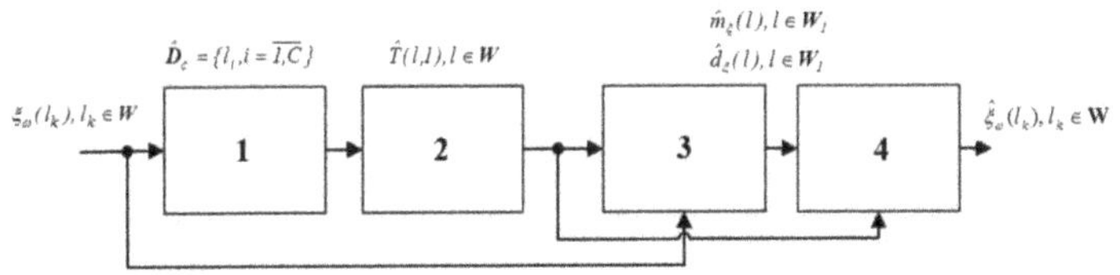

Figure 2. Block diagram of the computer modeling of the microrelief process (known approach): 1—assessment of the segmental structure of the microrelief; 2—assessment of the rhythmic structure of microrelief; 3—statistical processing of microrelief; 4—cyclic microrelief modeling.

In this case, the indicator functions, which allocate segments-cycles, were defined as:

$$I_{\mathbf{W}_i}(l) = \begin{cases} 1, & l \in \mathbf{W}_i, \\ 0, & l \notin \mathbf{W}_i. \end{cases}, \ i = \overline{1, C} \tag{3}$$

where $\mathbf{W}_i$ is the definition area of the indicator function, which in case of a discrete signal is $\mathbf{W} = \mathbf{D}$, that is, equals a discrete set of samples.

$$\mathbf{W}_i = \left\{ l_{i,j}, j = \overline{1, J} \right\}, \ i = \overline{1, C}, \tag{4}$$

The segmental cyclic structure $\hat{\mathbf{D}}_c$ is taken into account by a set of spatial samples $\{l_i\}$ or $\left\{l_{i,j}\right\}$, $i = \overline{1, C}$, $j = \overline{1, J}$. The mathematical model (1) takes into account the rhythm of the cyclic microrelief using the continuous rhythm function $T(l, n)$, namely:

$$I_{\mathbf{W}_i}(l) = I_{\mathbf{W}_{i+n}}(l + T(l, n)), \ i = \overline{1, C}, \ n = 1, \ l \in \mathbf{W}. \tag{5}$$

In order to assess the rhythm function $T(l, n)$, the segmental structure of the microrelief (in this case, the segmental cyclic structure) was determined as $\hat{\mathbf{D}}_c = \left\{ l_i, \ i = \overline{1, C} \right\}$, which is a set of spatial moments that correspond to the boundaries of the segments-cycles of the microrelief.

Having obtained the segmental structure $\hat{\mathbf{D}}_c$ and estimated the rhythmic structure (discrete rhythm function $T(l, n)$), we used the methods of statistical processing. As a result, we obtained statistical estimates of probabilistic characteristics (mathematical expectation $\hat{m}_\xi(l), \ l \in \mathbf{W}_1$ and dispersion $\hat{d}_\xi(l), \ l \in \mathbf{W}_1$) of the cyclic microrelief process. After this, the obtained information was used for computer simulation of the realization of the cyclic process of surface microrelief $\hat{\xi}_\omega(l_k), \ l_k \in \mathbf{W}$.

For an adequate description of the real process, the microrelief amplitudes on the segments-cycles need to be considered (this was prevented by the mathematical model presented in [14]); therefore, we take them into account in the new mathematical model, Figure 3.

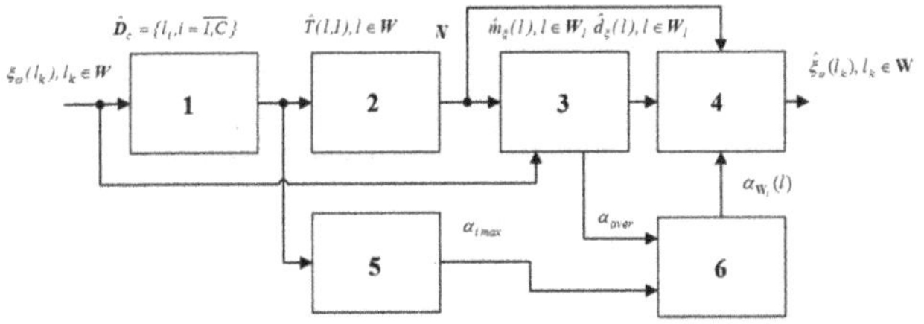

Figure 3. Block diagram of the computer modeling of the microrelief process (new approach): 1—assessment of the segmental structure of the microrelief; 2—assessment of the rhythmic structure of microrelief; 3—statistical processing of microrelief; 4—cyclic microrelief modeling; 5—determination of maximums of segments-cycles of microrelief; 6—estimation of the scale factors of the microrelief amplitude.

In the new model (1), segments-cycles of the cyclic microrelief process are defined as multiplicative components, taking into account the indicator functions and scale coefficients of the microrelief amplitude, that is:

$$f_i(l) = \xi_\omega(l) \cdot \alpha_{\mathbf{W}_i}(l) \cdot I_{\mathbf{W}_i}(l),\ i = \overline{1, C},\ l \in \mathbf{W}. \tag{6}$$

In Formula (6), an additional component $\alpha_{\mathbf{W}_i}(l)$ is introduced, which reflects the scale factors of the microrelief amplitude on each segment-cycle of the cyclic process, namely:

$$\alpha_{\mathbf{W}_i}(l) = \begin{cases} \alpha_i,\ l \in \mathbf{W}_i, \\ 0,\ l \notin \mathbf{W}_i. \end{cases},\ i = \overline{1, C}, \tag{7}$$

where α_i are the scale coefficients of the microrelief amplitude on each i-th segments-cycles defined as follows:

$$\alpha_i = \frac{\alpha_{i\max}}{\alpha_{aver}},\ i = \overline{1, C}, \tag{8}$$

where $\alpha_{i\max}$ is the maximum value of the microrelief amplitude on the i-th segment-cycle (determined at the segmentation stage of the cyclic microrelief process), α_{aver} is the average value of the microrelief amplitude (the maximum value of the mathematical expectation amplitude determined at the stage of statistical processing of the cyclic microrelief process).

4. Modeling Results

A comparative analysis of the results of computer modeling was performed. It showed that the new method of computer modeling of microrelief was more accurate than the well-known method presented in [14]. This is because the new method is based on a mathematical model of the self-ordered relief, which is presented as a cyclic random process, taking into account the amplitude parameters on each segment-cycle, Figure 4. Higher accuracy of computer modeling was achieved due to adapting the description of the shape of the segments-cycles under study.

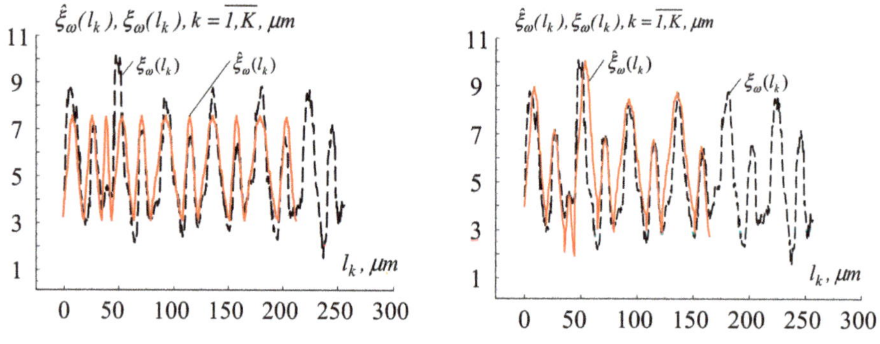

(a) (b)

Figure 4. The result of modeling the surface microrelief of titanium alloy Ti6Al4V after processing with femtosecond pulses: (**a**) based on the well-known mathematical model [14]; (**b**) based on the proposed mathematical model (red line—experimental data; black line—results of modeling).

This approach eliminated the negative effect of taking into account only the statistical evaluation (mathematical expectation) in case of modeling the relief morphology in the presence of a significant height variation of the microrelief elements.

Estimation of errors. In order to compare the proposed and well-known mathematical models, computer modeling of microrelief realizations was performed, and modeling results were estimated by defining the absolute $\Delta_q(k)$ and relative errors $\delta_q(k)$, Figure 5.

$$\Delta_q(k) = \sqrt{\frac{1}{K}\sum_{k=1}^{K}\left(\xi_\omega(l_k) - \hat{\xi}_\omega(l_k)\right)^2},\ q = \overline{1,2},\ k = \overline{1,K},\ l_k \in \mathbf{W}, \tag{9}$$

$$\delta_q(k) = \frac{\Delta_q(k)}{\sqrt{\frac{1}{K}\sum\limits_{k=1}^{K}\left(\hat{\xi}_\omega(l_k)\right)^2}},\ q = \overline{1,2},\ k = \overline{1,K},\ l_k \in \mathbf{W}. \tag{10}$$

where $\hat{\xi}_\omega(l_k)$ is the computer-simulated realization of the cyclic microrelief process (one of the two approaches); $\xi_\omega(l_k)$ is the experimentally obtained realization of the cyclic microrelief process; computer simulation errors based on the well-known model were identified at q = 1; computer simulation errors based on the new model were identified at q = 2 (new approach).

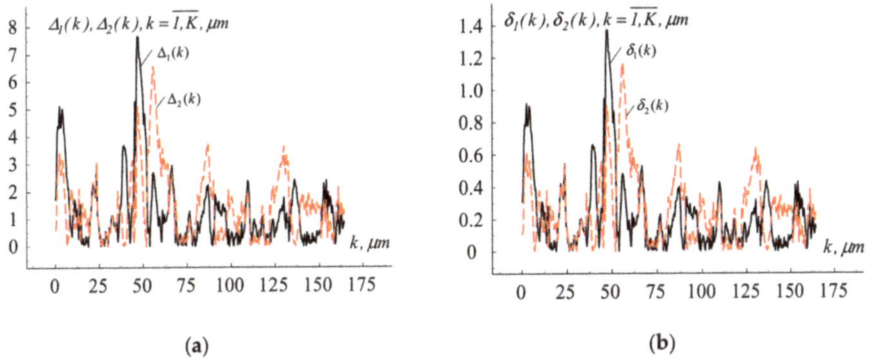

Figure 5. Fragments of the absolute and relative errors of computer modeling of the surface microrelief of titanium alloy Ti6Al4V after processing with femtosecond pulses: (**a**) absolute errors, (**b**) relative errors.

It was found that the relative root mean square modeling error for the studied case (new approach) did not exceed 1.2 (12%). It should be noted that the comparative analysis of the relief reproduction accuracy was individual for each realization of the relief formation process, Table 1.

Table 1. Comparison of characteristics of the well-known and new mathematical models of the surface relief after laser treatment.

Models	Taking into Account the Cyclical Nature of the Relief	Taking into Account the Random Nature of the Relief	Taking into Account the Morphological Features of Segments-Cycles of the Relief	Taking into Account the Rhythmic Features of the Deployment of Segments-Cycles of the Relief	Taking into Account the Amplitude Features of Segments-Cycles of the Relief
Well-known [5–7,16–19]	+	+	+/−	+	−
New	+	+	+	+	+

"+"—takes into account (reflects); "−"—does not take into account (does not reflect); "+/−"—partially takes into account (partially reflects).

Based on a new mathematical model of the surface relief, as well as previously developed approaches to estimating the segmental and rhythmic structures and probabilistic characteristics of the process [19], a new approach to computer modeling of the cyclic microrelief realization on implant surfaces has been substantiated [20,21].

5. Conclusions

Based on the new mathematical model of surface microrelief, as applied to titanium alloy Ti6Al4V after its processing with femtosecond pulses, a method of computer modeling of the surface microrelief realization was developed. The model contains the components that take into account changes in the shape of cycles-segments of the processes investigated. The developed method of computer modeling makes it possible to describe more precisely the features of microrelief segments. This allowed increasing the efficiency of their processing procedure and computer modeling in information systems.

The software for computer simulation of the surface microrelief realization created on the basis of the new model can be integrated into specialized software for technical diagnostics of the surface condition and modeling experiments conducted after the precision laser processing. Using the developed software, a series of experiments on the processing of real microrelief sections by the new and well-known methods was performed. The obtained results of comparative analysis of modeling errors using the new method confirmed its higher accuracy in describing the segments-cycles as compared to the well-known method.

In further scientific research, it is planned to modify the mathematical model so as to allow considering the features of the microrelief nonlinearity (presence of a trend) on curvilinear surfaces.

Author Contributions: Conceptualization, I.L., V.H., G.S.; formal analysis, P.M. and V.D.; investigation, V.D., P.M., V.H., I.L., G.S.; methodology, I.L. and P.M.; project administration, P.M., H.S.; validation, I.L., P.M.; writing—original draft, I.L., P.M., V.H., V.D.; writing—review and editing, V.H., H.S. All authors have read and agreed to the published version of the manuscript.

Funding: This research received no external funding.

Conflicts of Interest: The authors declare no conflict of interest.

References

1. Van Driel, H.M.; Sipe, J.E.; Young, J.F. Laser-induced periodic surface structure on solids: A universal phenomenon. *Phys. Rev. Lett.* **1982**, *49*, 1955–1958. [CrossRef]
2. Vorobyev, A.; Guo, C. Femtosecond laser structuring of titanium implants. *Appl. Surf. Sci.* **2007**, *253*, 7272–7280. [CrossRef]
3. Kuczyńska-Zemła, D.; Kijeńska-Gawrońska, E.; Pisarek, M.; Borowicz, P.; Swieszkowski, W.; Garbacz, H. Effect of laser functionalization of titanium on bioactivity and biological response. *Appl. Surf. Sci.* **2020**, *525*, 146492. [CrossRef]
4. Schnell, G.; Duenow, U.; Seitz, H. Effect of laser pulse overlap and scanning line overlap on femtosecond laser-structured Ti6Al4V surfaces. *Materials* **2020**, *13*, 969. [CrossRef] [PubMed]
5. Filimonova, E.A. Development of the Methodology and Program of the Automated Control of Microgeometry of Surfaces of Details of Instruments Using Graphic Criteria and Their Use in Technological Research, C. Ph.D. Dissertation, ITMO University, Saint Petersburg, Russia, 2014; p. 237. (In Russian).
6. Gibadullin, I.N.; Valetov, V.A. Image of the surface profile as a graphic criterion of its roughness. *J. Instrum. Eng.* **2019**, *62*, 86–92. (In Russian) [CrossRef]
7. Zhou, W.; Tang, J.Y.; He, Y.F.; Zhu, C.C. Modeling of rough surfaces with given roughness parameters. *J. Cent. South Univ.* **2017**, *24*, 127–136. [CrossRef]
8. Vorobyev, A.Y.; Guo, C. Direct femtosecond laser surface nano/microstructuring and its applications. *Laser Photon. Rev.* **2012**, *7*, 1–23. [CrossRef]
9. Kuznetsov, G.V.; Feoktistov, D.V.; Orlova, E.G.; Batishcheva, K.; Ilenok, S.S. Unification of the textures formed on aluminum after laser treatment. *Appl. Surf. Sci.* **2019**, *469*, 974–982. [CrossRef]

10. Schnell, G.; Polley, C.; Bartling, S.; Seitz, H. Effect of chemical solvents on the wetting behavior over time of femtosecond laser structured Ti6Al4V surfaces. *Nanomaterials* **2020**, *10*, 1241. [CrossRef] [PubMed]
11. Varlamova, O.; Reif, J.; Varlamov, S.; Bestehorn, M. Self-organized Surface Patterns Originating from Laser-Induced Instability. In *Progress in Nonlinear Nano-Optics. Nano-Optics and Nanophotonics*; Sakabe, S., Lienau, C., Grunwald, R., Eds.; Springer: Cham, Switzerland, 2015. [CrossRef]
12. Romano, J.-M.; Garcia-Giron, A.; Penchev, P.; Dimov, S. Triangular laser-induced submicron textures for functionalising stainless steel surfaces. *Appl. Surf. Sci.* **2018**, *440*, 162–169. [CrossRef]
13. Aguilar-Morales, A.I.; Alamri, S.; Kunze, T.; Lasagni, A.F. Influence of processing parameters on surface texture homogeneity using direct laser interference patterning. *Opt. Laser Technol.* **2018**, *107*, 216–227. [CrossRef]
14. Lytvynenko, I.V.; Maruschak, P.O. Analysis of the state of the modified nanotitanium surface with the use of the mathematical model of a cyclic random process. *Optoelectron. Instrument. Proc.* **2015**, *51*, 254–263. [CrossRef]
15. Lytvynenko, I.V.; Maruschak, P.O.; Lupenko, S.A.; Popovych, P.V. Modeling of the ordered surface topography of statically deformed aluminum alloy. *Mater. Sci.* **2016**, *52*, 113–122. [CrossRef]
16. Frischer, R.; Krejcar, O.; Selamat, A.; Kuca, K. 3D surface profile diagnosis using digital image processing for laboratory use. *J. Cent. South Univ.* **2020**, *27*, 811–823. [CrossRef]
17. Wang, X.; Feng, C. Development of empirical models for surface roughness prediction in finish turning. *Int. J. Adv. Manuf. Technol.* **2002**, *20*, 348–356. [CrossRef]
18. Ozcelik, B.; Bayramoglu, M. The statistical modeling of surface roughness in high-speed flat end milling. *Int. J. Mach. Tools Manuf.* **2006**, *46*, 1395–1402. [CrossRef]
19. Lytvynenko, I.V.; Maruschak, P.O.; Lupenko, S.A.; Hats, Y.I.; Menou, A.; Panin, S.V. Software for segmentation, statistical analysis and modeling of surface ordered structures. In *AIP Conference Proceedings*; AIP Publishing LLC: Melville, NY, USA, 2016; Volume 1785, p. 030012. [CrossRef]
20. Muller, F.; Kunz, C.; Graf, S. Bio-inspired functional surfaces based on laser-induced periodic surface structures. *Materials* **2016**, *9*, 476. [CrossRef] [PubMed]
21. Schnell, G.; Staehlke, S.; Duenow, U.; Nebe, J.B.; Seitz, H. Femtosecond laser nano/micro textured Ti6Al4V surfaces—Effect on wetting and MG-63 cell adhesion. *Materials* **2019**, *12*, 2210. [CrossRef] [PubMed]

materials MDPI

Article

The Multi-Stage Drawing Process of Zinc-Coated Medium-Carbon Steel Wires in Conventional and Hydrodynamic Dies

Maciej Suliga [1,*], Radosław Wartacz [1] and Marek Hawryluk [2]

[1] Faculty of Production Engineering and Materials Technology, Czestochowa University of Technology, 19 Armii Krajowej Av. 42-200 Czestochowa, Poland; imitm@wip.pcz.pl

[2] Department of Metal Forming and Metrology, Wroclaw University of Science and Technology, 5 Lukasiewicza str., 50-371 Wrocław, Poland; marek.hawryluk@pwr.edu.pl

* Correspondence: maciej.suliga@pcz.pl; Tel.: +48-343250786

Received: 6 October 2020; Accepted: 27 October 2020; Published: 30 October 2020

check for updates

Abstract: This paper discusses experimental studies aiming to determine the effect of the drawing method on the lubrication conditions, zinc coating mass and mechanical properties of medium-carbon steel wires. The test material was 5.5 mm-diameter galvanized wire rod which was drawn into 2.2 mm-diameter wire in seven draws at a drawing speed of 5, 10, 15, 20 and 20 m/s, respectively. Conventional and hydrodynamic dies with a working portion angle of $\alpha = 5°$ were used for the drawing process. It has been shown that using hydrodynamic dies in the process of multi-stage drawing of zinc-coated wire improves the lubrication conditions, which leads to a reduction in friction at the wire/die interface. As a consequence, wires drawn hydrodynamically, as compared to wires drawn conventionally, are distinguished by a thicker zinc coating and better mechanical and technological properties.

Keywords: wire; zinc coating; hydrodynamic die; drawing speed; surface; properties

1. Introduction

External friction between the wire and the drawing die is among the factors which determines the conditions of the top wire layer [1–5]. Among the known methods of improving lubrication conditions and reducing friction is the application of hydrodynamic dies, the so-called pressure dies in the drawing process. The drawing process performed with the use of hydrodynamic dies consists in introducing a lubricant into the resistance die through a narrow slot. As a result, the lubricant becomes a liquid, successively increasing the pressure in the die, where the greatest value is achieved at the contact of the working die with the wire. The generated high pressure not only separates the friction surfaces of the wire and the working die but also contributes to the deformation by changing the diameter and shape of the drawn wire [6–8].

In the case of drawing zinc-coated wire, this layer consists of zinc coatings and a steel substrate [9]. In hot-dip zinc coating, the properties of the zinc coating depend largely on the galvanizing technology, including the composition and temperature of the bath and the immersion time [9–12]. In contrast to drawing uncoated wire, a very high variation in physicochemical properties between individual wire layers exists in the surface layer of zinc-coated wire at a depth of approximately 100 μm. In the multi-stage drawing process, the top surface layer is intensively heated up. This, in turn, is expected to influence the conditions of deformation of the soft zinc coating on the hard steel core, including wire surface roughness. The formation of the wire topography also depends on the drawing technology. The literature on the zinc-coated wire drawing technology lacks information concerning the surface roughness of zinc-coated wire after the multi-stage drawing process at high drawing speeds (above 5 m/s). Nevertheless, the literature data on single-stage drawing at a speed of up to 2 m/s show that the angle and the drawing method influence both the top wire layer and the zinc coating surface roughness [13–16].

2. Materials and Methods

The test material was 5.5 mm-diameter galvanized steel wire rod (Drumet, Wloclawek, Poland). The examination of the microstructure of the coatings was made with the use of an S-3400 N-type Hitachi scanning microscope equipped (Hitachi, Tokyo, Japan) with an Energy Dispersion Spectroscopy X-ray spectrometer. An accelerating voltage of 25 kV was used. Figure 1 shows the structure of the zinc coating before and after the drawing process (wire rod with a diameter of 5.5 mm).

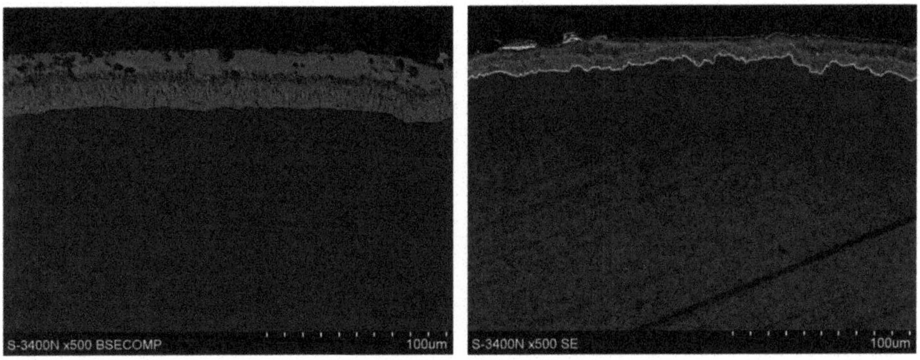

Figure 1. Coat cross-section (SEM) microstructure for: (**a**) wire rod, (**b**) 2.2 mm wire drawn in conventional dies.

The SEM investigation showed that after the hot-dip galvanizing process of the wire rod, the zinc coating consists of an outer layer and a diffusion layer containing intermetallic phase layers, Figure 1a. The process of drawing wires with a diameter of 2.2 mm caused, regardless of the drawing technology, a more than two-fold decrease in the thickness of the zinc coating, Figure 1b.

The wire rod was drawn into 2.2 mm-diameter wire on a multi-stage drawing machine in seven draws at a drawing speed of 5, 10, 15, 20 and 20 m/s, respectively (Tables 1 and 2). Conventional dies and hydrodynamic dies with a working portion angle of $\alpha = 5°$ were used for the drawing process.

Table 1. Distribution of single reductions, G_p, and the total reduction, G_c.

Draw No.	0	1	2	3	4	5	6	7
ϕ, mm	5.50	4.73	4.10	3.57	3.13	2.77	2.46	2.20
G_p, %	-	26.04	24.86	24.18	23.13	21.68	21.13	20.02
G_c, %	-	26.04	44.43	57.87	67.61	74.64	79.99	84.00

The examination of the surface topography of the zinc-coated wires after the drawing process was carried out on a Form Talysurf 50e profilometer (Taylor Hobson, Leicester, England). Figure 2 shows an example of a profilogram of the surface texture of 2.20 mm-diameter wire.

Table 2. Drawing speed, v [m/s], at individual drawing drafts.

Draw No.						
1	2	3	4	5	6	7
Drawing speed, v [m/s], at individual drafts						
1.06	1.43	1.90	2.47	3.15	4.00	5
2.12	2.86	3.80	4.94	6.31	8.00	10
3.17	4.28	5.70	7.41	9.46	12.00	15
4.22	5.70	7.59	9.88	12.62	16.00	20

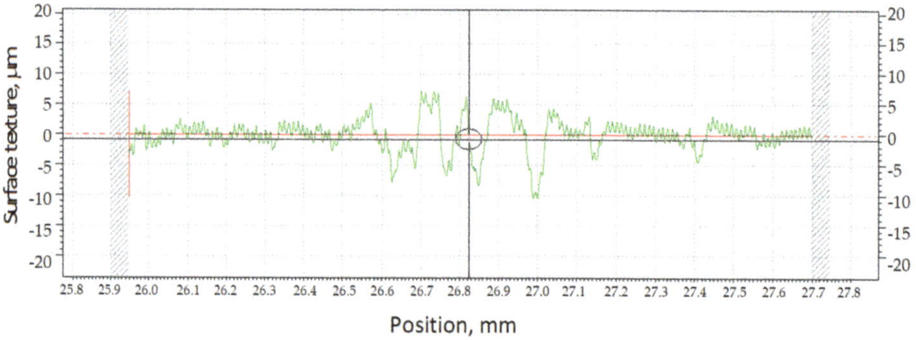

Figure 2. The surface texture of ϕ2.2 mm-diameter wire drawn in hydrodynamic dies.

Figure 2 shows an example of a profilogram of the surface texture of 2.20 mm-diameter wire. To illustrate the variations in the wire surface texture, the profile height and profile deviation parameters were used for the analysis. The average line (Figure 2) is understood as a profilogram line which divides the surface into two parts in such a manner that the surface calculated using the integral is equal to zero. The circle in the center of the profilogram of the geometric structure of the wire surface represents the center of the wire roughness measurement on the analyzed length, while the numbers on the x-axis represent the current position of the measuring head of the device.

The profile height parameter describes the height of the profile irregularities using linear dimensions perpendicular to the average line or the arithmetic mean. The explanation of the profile height can be represented with the use of the following parameters:

R_p—the maximum height of the profile elevation above the value of the average line within the measuring length under examination;

R_v—the maximum depth of the profile below the average line within the measuring length;

R_t—the total value of the profile depth and profile height within the measuring length under examination:

$$R_t = R_p + |R_v| \quad (1)$$

R_z—the arithmetic mean of the absolute values of the heights of the five largest elevations and the five largest depressions of the roughness profile:

$$R_z = \frac{\sum_{i=1}^{5}\left|y_{pi}\right| + \sum_{i=1}^{5}\left|y_{ri}\right|}{5} \quad (2)$$

The profile deviation parameters describe the deviation of the profile in the direction perpendicular to the average line, where the basis for the calculation of the measure of dispersion is the assumption of mathematical statistics. They are calculated as follows:

R_a—the mean of the profile deviations from the average line:

$$R_a = \frac{1}{L}\int_0^L |y(x)|dx \quad (3)$$

R_q—the quadratic mean of the values of profile deviations from the average line within the examined measuring length:

$$R_q = \sqrt{\frac{1}{1}\int_0^1 y^2(x)dx} \quad (4)$$

The examination of the zinc-coated wire roughness measurements was carried out for finished 2.2 mm-diameter wires. For the structural examination of the top wire layer geometry, five elementary segments, each of a measuring length of 0.8 mm, were used. The wire roughness measurements were performed for the longitudinal direction to the drawing direction.

3. Results and Discussion

Lubrication conditions. To determine the effect of the drawing method on the lubrication conditions, ten 100 mm-long specimens were taken for each drawing variant. After the specimens were weighed on a laboratory balance, the lubricant layer was removed with the use of sodium hydroxide (NaOH) and technical acetone. After the specimens had completely dried up, they were weighed again. From the mass difference, the amount of lubricant on the wire surface was determined. The test results are presented in Figures 3 and 4 for conventional dies (K) and hydrodynamic dies (H), respectively.

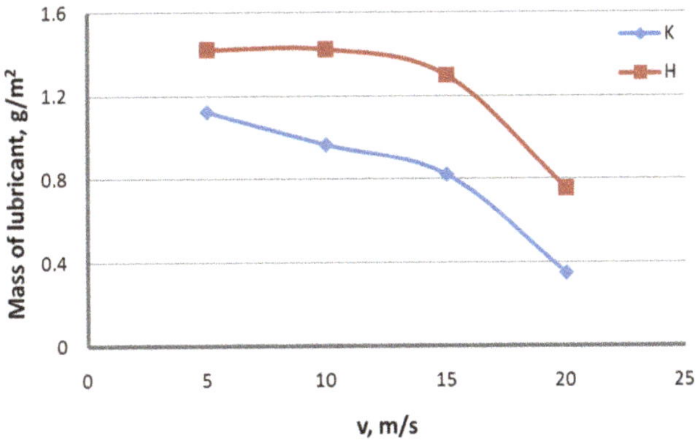

Figure 3. The effect of the drawing speed, v, on the lubricant mass on the surface of 2.2 mm-diameter wires drawn in conventional dies (K) and hydrodynamic dies (H), respectively.

The test results illustrated in Figures 2 and 3 confirm the significant effect of the drawing method on the lubrication conditions. The data in Figures 2 and 3 show that using hydrodynamic dies in the zinc-coated wire drawing process improves the lubrication conditions, while the higher the drawing speed and the greater the total reduction, the larger the differences. For wires drawn hydrodynamically at drawing speeds of 5 and 20 m/s, as compared to wires drawn conventionally, a lubricant amount larger by 26.6% and 114.8%, respectively, was recorded on the wire surface. The improvement in the lubrication conditions when drawing in hydrodynamic dies can be explained by more favorable conditions for the deformation of the top layer of the zinc-coated wire.

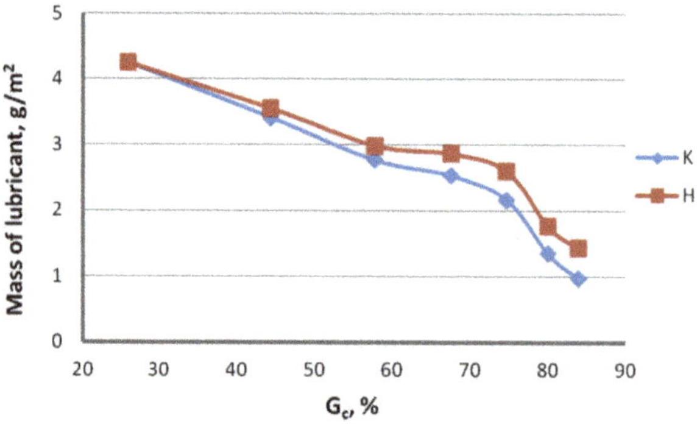

Figure 4. Variation in the lubricant mass on the surface of wires drawn in conventional dies (K) and hydrodynamic dies (H) as a function of total reduction (v = 10 m/s).

Analysis of the surface topography of zinc-coated wires. In drawing processes, aside from the die geometry and the drawing speed, the method of drawing has a significant influence on the formation of the wire surface. Therefore, a comparison of the surface roughness of zinc-coated wires drawn in conventional dies and hydrodynamic dies, respectively, was made in the study.

Based on the results, graphs were plotted to illustrate the effect of the drawing method on the surface texture parameters of 2.2 mm-diameter zinc-coated wires—see Figures 5–10.

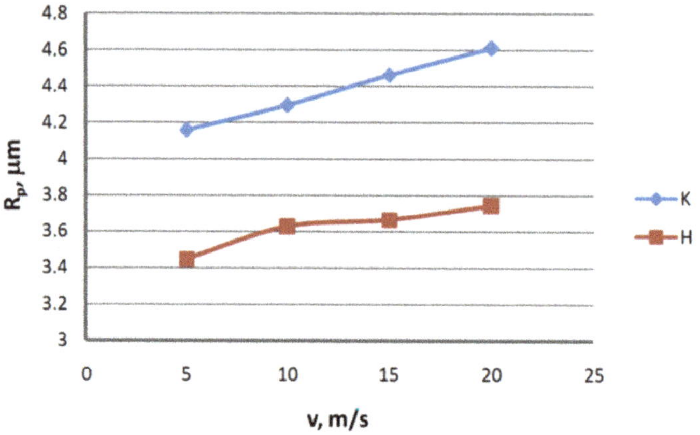

Figure 5. The effect of the drawing speed, v, on the surface profile height parameter, R_p, for 2.2 mm-diameter wires drawn in conventional dies (K) and hydrodynamic dies (H), respectively.

The analysis of the surface roughness of wires drawn conventionally and hydrodynamically has shown that using hydrodynamic drawing dies reduces the surface roughness of zinc-coated wire. At a drawing speed of 20 m/s, the difference between the drawing variants is as follows: 12.5% R_v, 20.4% R_v, 15.8% R_t, 11.3% R_z, 18.7% R_a and 44.6% R_q, respectively. Using hydrodynamic dies in the drawing process creates more favorable conditions for the deformation of the top layer of the zinc-coated wire. The better lubrication conditions, as confirmed by a greater amount of the lubricant after the drawing process (Figure 3), have contributed to a reduction in friction and the wire top surface temperature. In contrast to drawing uncoated wire, significant changes to the physicochemical properties of the zinc coating occur in the case of drawing zinc-coated wire. Using hydrodynamic dies diminishes the adverse effect of the drawing speed on the softening of the thin zinc coating in the die. Hence, wires drawn by the hydrodynamic method have a smoother surface.

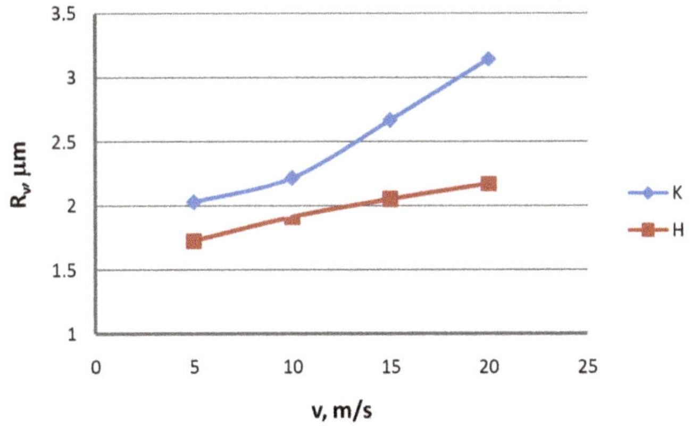

Figure 6. The effect of the drawing speed, v, on the surface profile height parameter, R_v, for 2.2 mm-diameter wires drawn in conventional dies (K) and hydrodynamic dies (H), respectively.

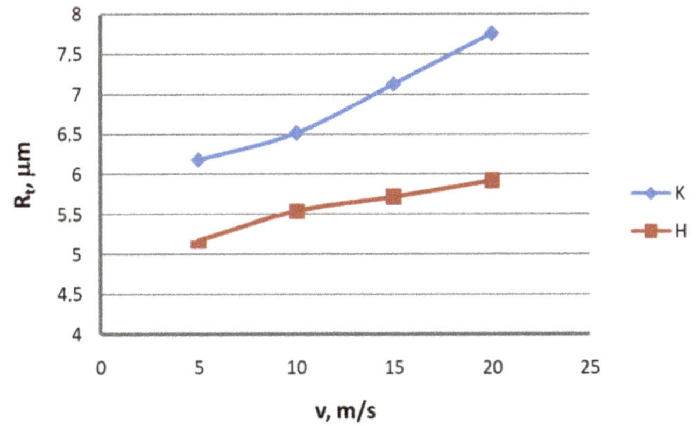

Figure 7. The effect of the drawing speed, v, on the surface profile height parameter, R_t, for 2.2 mm-diameter wires drawn in conventional dies (K) and hydrodynamic dies (H), respectively.

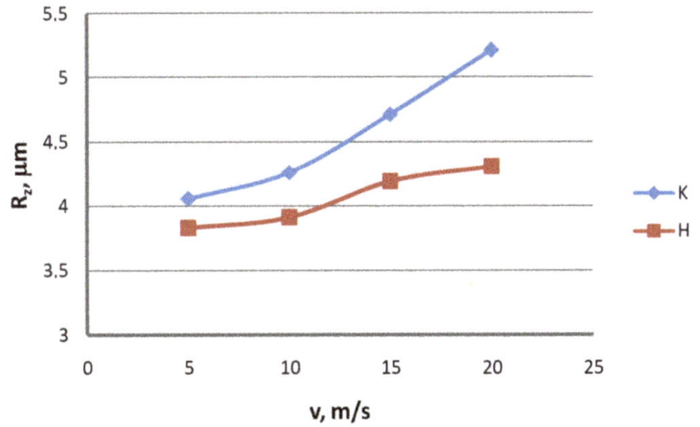

Figure 8. The effect of the drawing speed, v, on the surface profile height parameter, R_z, for 2.2 mm-diameter wires drawn in conventional dies (K) and hydrodynamic dies (H), respectively.

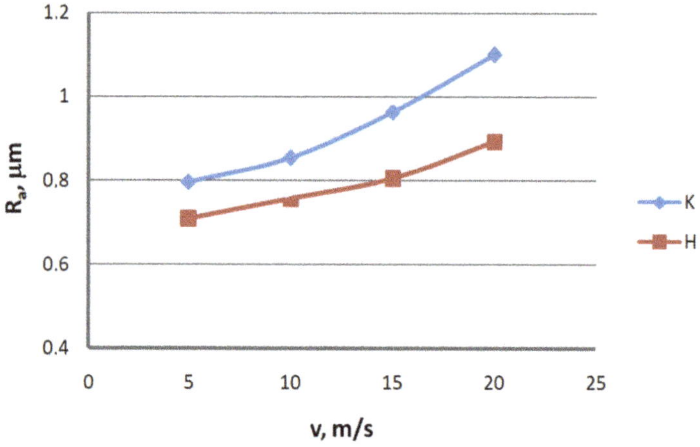

Figure 9. The effect of the drawing speed, v, on the surface profile deviation parameter, R_a, for 2.2 mm-diameter wires drawn in conventional dies (K) and hydrodynamic dies (H), respectively.

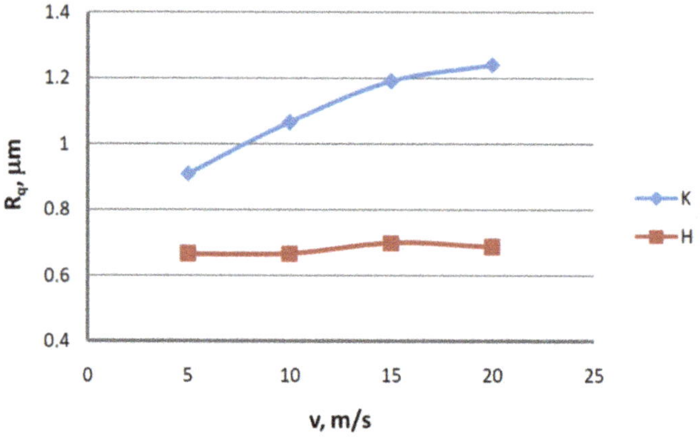

Figure 10. The effect of the drawing speed, v, on the surface profile deviation parameter, Rq, for 2.2 mm-diameter wires drawn in conventional dies (K) and hydrodynamic dies (H), respectively.

Examination of zinc coating mass. The mass of the zinc coating was determined by the gravimetric method in accordance with the applicable standard PN-EN 10244-1. Then, the change in the zinc coating mass after drawing in conventional dies and hydrodynamic dies was compared. The test results are illustrated in Figures 11 and 12.

The analysis of the surface roughness of wires drawn conventionally and hydrodynamically has shown that using hydrodynamic drawing dies reduces the surface. It can be seen from Figures 11 and 12 that the drawing method significantly influences the variation of the zinc coating mass on the wire after the drawing process. Using hydrodynamic dies in the multi-stage zinc-coated wire drawing process favorably influences the drawing conditions and the zinc coating mass. While at a drawing speed of 5 m/s, the difference in the zinc mass between the wires drawn conventionally and hydrodynamically amounts to 3.8%, at a drawing speed of 20 m/s, it already exceeds 47%. It has also been found that the difference in the zinc mass between variants K and H increases with the increase in total reduction. At a reduction of Gc = 84% and a speed of 10 m/s, this difference is approximately 15%.

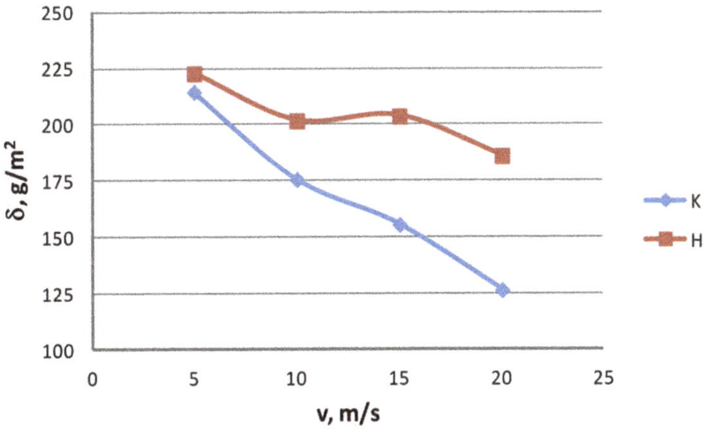

Figure 11. The effect of the drawing speed, v, on the zinc mass δ on the surface of 2.2 mm-diameter wires drawn in conventional dies (K) and hydrodynamic dies (H), respectively.

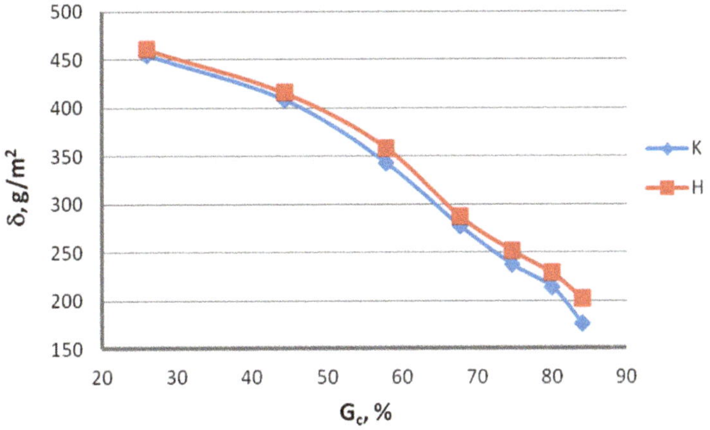

Figure 12. Variation in the zinc mass δ on the surface of wires drawn in conventional dies (K) and hydrodynamic dies (H) as a function of total reduction (v = 10 m/s).

When analyzing the mass of the coating (based on standard PN-EN 10244-2), which should be present on the wire when it is categorized into the respective class, it can be found that, for a drawing speed of 5 m/s, the coatings on wires drawn either conventionally or hydrodynamically are situated in class AB, whereas the coating in the hydrodynamic method is thicker by 3.8% compared to that in the conventional method. At a drawing speed of 10 m/s, both variants still hold class AB, while the difference between the variants being almost 15% of the zinc mass on the other wire.

A significant difference of 30% in the zinc coating mass between the variants under analysis caused variant K (conventional dies) to be categorized into a lower class—i.e., B, at a drawing speed of 15 m/s, while variant H (hydrodynamic dies) achieved class AB. The increase in the drawing speed from 15 to 20 m/s did not change the class of the coatings.

In the hydrodynamic method, the mass of the remaining zinc coating is dependent on the lubricant pressure created in the die during the drawing process, which directly influences the friction conditions and the wire heating. Sufficiently high lubricant pressure in the hydrodynamic die allowed relatively good lubrication conditions to be maintained. This, in turn, contributed to a lowering of wire temperature. In consequence, the wires drawn hydrodynamically were distinguished by a thicker zinc coating. Unlike the conventional method, no such negative effect of high drawing speeds on the zinc coating thickness was noted for the hydrodynamic method. Therefore, the wires drawn in hydrodynamic dies at a drawing speed of 20 m/s had the coating class AB.

Testing for mechanical and engineering properties. Tests aiming to determine the mechanical and engineering properties of the wire were performed in accordance with standard PN-EN 10218-1:2012 on a Zwick/Z100 testing machine and on a wire twisting and bending test device. Wires of a diameter of 2.2 mm were subjected to testing to determine the yield strength, $R_{0.2}$; ultimate tensile strength, R_m; uniform elongation, A_r; total elongation, A_c; reduction in area, Z; the number of twists, N_t; and the number of bends, N_b. The results of the mechanical and technological tests are represented in Figures 13–17. Figure 13 shows that using hydrodynamic dies results in a reduction in the mechanical properties of zinc-coated steel wire. Wires drawn in hydrodynamic dies, as compared to wires drawn conventionally, showed a yield point decrease, on average, of 2.7% and an ultimate tensile strength decrease of 2.3%.

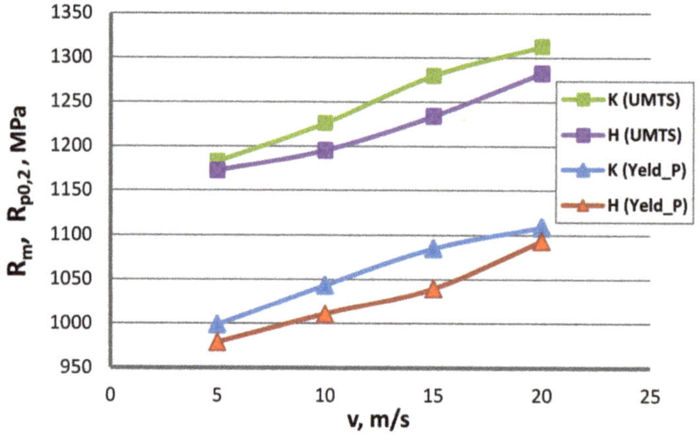

Figure 13. The effect of the drawing speed, v, on the yield strength (UMTS) and yield point, $R_{0.2}$, of 2.2 mm-diameter wires drawn in conventional dies (K) and hydrodynamic dies (H), respectively.

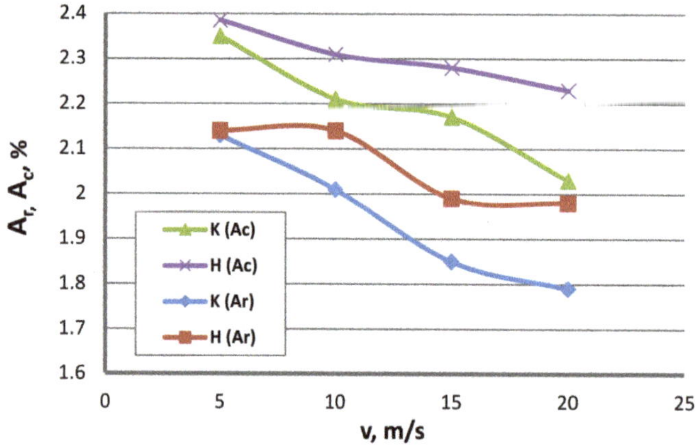

Figure 14. The effect of the drawing speed, v, on the uniform elongation, A_r and also on the total elongation Ac of 2.2 mm-diameter wires drawn in conventional dies (K) and hydrodynamic dies (H), respectively.

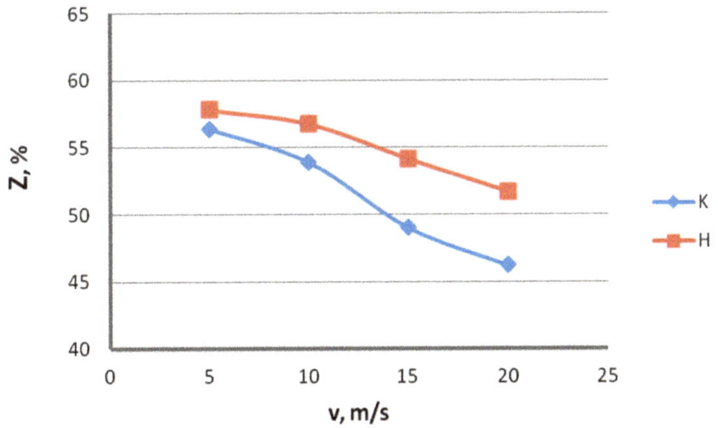

Figure 15. The effect of the drawing speed, v, on the reduction in area, Z, of 2.2 mm-diameter wires drawn in conventional dies (K) and hydrodynamic dies (H), respectively.

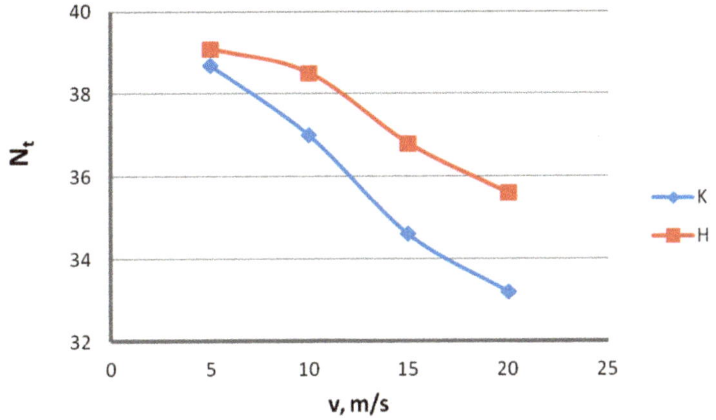

Figure 16. The effect of the drawing speed, v, on the number of twists, N_t, of 2.2 mm-diameter wires drawn in conventional dies (K) and hydrodynamic dies (H), respectively.

The lower strain hardening of the hydrodynamically drawn wires can be linked with more favorable conditions of lubrication as well as wire and zinc coating deformation in the drawing process. Hence, the wires drawn by the hydrodynamic method exhibited better plastic properties, as confirmed by the data illustrated in Figures 14 and 15.

The data in Figures 14 and 15 show that, with the increase in the drawing speed, the differences in the plastic properties between the conventionally and hydrodynamically drawn wires increase and exceed 10% at a drawing speed of 20 m/s. Wires drawn hydrodynamically at a drawing speed of 20 m/s, as compared to wires drawn conventionally, show reductions in uniform elongation and total elongation values of 10.6% and 9.9%, respectively, as well as a reduction in the area of 11.8%. Using hydrodynamic dies in the zinc-coated wire drawing process also has a favorable effect on the technological properties, such as the number of twists, N_t, and the number of bends, N_b. This is confirmed by the data shown in Figures 16 and 17.

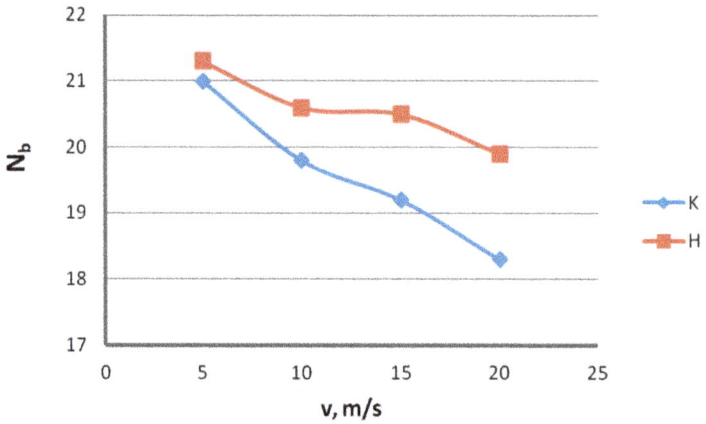

Figure 17. The effect of the drawing speed, v, on the number of bends, N_b, of 2.2 mm-diameter wires drawn in conventional dies (K) and hydrodynamic dies (H), respectively.

Using hydrodynamic dies largely removes the adverse effect of the drawing speed on the technological properties of the wire. In the drawing speed range of 5–20 m/s, using hydrodynamic dies caused an increase in the number of twists, N_t, by 1% to 7.2% and an increase in the number of bends, N_b, by 1.4% to 8.7%, compared to conventional dies. Therefore, the use of hydrodynamic dies in the zinc-coated wire drawing process enables not only a thicker zinc coating on the finished wire but also better mechanical and technological properties to be obtained, especially at drawing speeds not exceeding 15 m/s.

Analysis of residual stresses in steel wires. Residual stress is one of the basic parameters significantly affecting the quality of the steel wire. Mechanical methods are the best methods for determining the residual stresses in wires, allowing quick stress measurements. The analysis of the influence of the drawing method on the residual stresses in end wires with a nominal diameter of 2.2 mm was determined based on the method of longitudinal wire cutting, known in the literature as the Schepers–Peiter method. This method involves cutting the wire to a certain length and measuring the deflection value at the end of the wire. The end wires with a nominal diameter of 2.2 mm were cut on a wire EDM machine to the length l = 44 mm (ratio l/d = 20). Schepers–Peiter [16] proposed that from the equation of the moments of forces acting on the cut wire, the following expression is obtained for the circular-symmetric distribution to determine the stresses on the outer surface of the wire:

$$\sigma_W = 1.3176 \cdot E \cdot R \cdot \frac{h}{l^2} \tag{5}$$

where:

σ_W—longitudinal residual stress on the wire surface, MPa; E—Yung's module, MPa; R—wire radius, mm; h—parting of the wire ends, mm; l—cutting length, mm.

The results of type I residual stresses (average values of 10 wires) are presented in Table 3. In order to better analyze the obtained test results, the percentage differences between the analyzed drawing variants are also presented.

Table 3. Results of the residual stress tests performed by the longitudinal cutting method for the final wires with a nominal diameter of 2.2 mm drawn in conventional dies (K) and in hydrodynamic dies (H).

v, m/s	Die Angle α, ° K	H	(K-H)/K, %
	Residual Stress σ_W, MPa		
5	290.6	167.1	42.5
10	312.7	189.0	39.6
15	349.4	252.7	27.7
20	383.3	370.6	3.3

On the basis of the results presented in Table 3, diagrams were prepared showing the influence of the drawing method on the first type residual stresses, Figure 18.

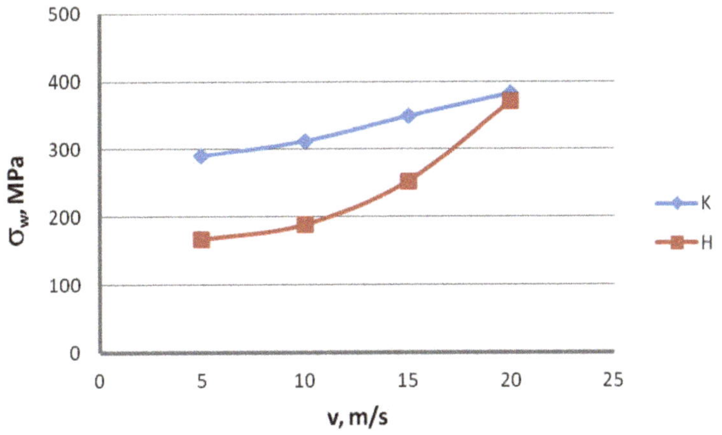

Figure 18. The effect of the drawing speed, v, on the longitudinal residual stresses σ_W of 2.2 mm-diameter wires drawn in conventional dies (K) and hydrodynamic dies (H), respectively.

Based on the data presented in Figure 18, it can be concluded that the drawing method significantly affects the residual stresses in steel wires. The use of hydrodynamic dies in the drawing of galvanized steel wires causes a decrease in residual stress, especially in the speed range of 5–15 m/s. With the increase in the drawing speed, the differences in the values of longitudinal residual stresses between the wires drawn with the conventional and hydrodynamic methods decreased, and were: 42.5%, 39.6%, 27.7% and 3.3%, respectively

Lower values of residual stresses in hydrodynamically drawn wires should be seen with the occurrence of smaller deformations for this variant. Separating the friction surfaces of the wire and die with a sufficiently thick layer of lubricant allowed a significant decrease in the friction coefficient, and this, in turn, reduced the deformation resistance. As a consequence, the wires drawn with the hydrodynamic method are characterized by a lower heterogeneity of the deformation on the wire cross-section, hence the decrease in residual stresses for this drawing variant. Lower residual stresses in hydrodynamically drawn wires can also be seen in a more uniform temperature distribution, because the lower the friction coefficient, the lower the temperature of the wire surface layer heated by friction.

4. Conclusions

The carrying out of the process of drawing zinc-coated wire in hydrodynamic dies enables an improvement in the lubrication conditions and reduces the friction at the wire/die interface. Therefore, a larger amount of zinc has been applied on the wire surface for wire drawn hydrodynamically, as opposed to wire drawn conventionally.

Using hydrodynamic dies partially removes the adverse effect of high drawing speeds (above 10 m/s) on the softening of the thin zinc coating in the die. In consequence, wire drawn by the hydrodynamic method has a smoother surface.

Wires drawn in hydrodynamic dies in the drawing speed range of 5-20 m/s showed, on average, a yield point decrease of 2.7% and an ultimate tensile strength decrease of 2.3%, as compared to wires drawn conventionally. The lesser strain hardening of hydrodynamically drawn wires can be associated with more favorable conditions of lubrication as well as wire and zinc coating deformation in the drawing process. Therefore, wires drawn by the hydrodynamic method were distinguished by better plastic properties.

Using hydrodynamic dies largely removes the adverse effect of the high drawing speeds on the engineering properties of the wire. Therefore, hydrodynamically drawn wires showed a number of twists greater by 1% to 7.2% and a number of bends greater by 1.4% to 8.7%.

The application of hydrodynamic dies in the drawing process of galvanized steel wires causes a decrease in residual stress, especially in the speed range 5–15 m/s. The separation of the friction surfaces of the wire and the hydrodynamic drawing die with a sufficiently thick layer of lubricant allowed a significant decrease in the friction coefficient, deformation resistance and deformation heterogeneity on the wire cross-section. In consequence, the wires drawn hydrodynamically are characterized by lower residual stresses.

The obtained investigation results can be used in the design of technologies for multi-stage drawing of zinc-coated steel wire. According to the authors, in the process of drawing galvanized wires, hydrodynamic dies should be used in all drafts.

Author Contributions: Conceptualization, M.S. and R.W.; methodology, M.S.; validation, M.S. and M.H.; formal analysis, M.S. and M.H.; investigation, M.S. and R.W.; resources, M.S.; data curation, M.S. and R.W.; writing—original draft preparation, M.S. and R.W.; writing—review and editing, M.H. and M.S.; visualization, R.W.; supervision, M.H.; project administration, M.S. All authors have read and agreed to the published version of the manuscript.

Funding: This research received no external funding.

Conflicts of Interest: The authors declare no conflict of interest. The funders had no role in the design of the study; in the collection, analyses, or interpretation of data; in the writing of the manuscript, or in the decision to publish the results.

References

1. Wartacz, R. Theoretical and Experimental Analysis of the Multistage Drawing of Galvanized Wires from C42D Steel. Ph.D. Thesis, Czestochowa University of Technology, Częstochowa, Poland, 17 June 2019.
2. Łuksza, J. *Drawing Elements*; AGH: Kraków, Poland, 2001.
3. Masse, T.; Fourment, L.; Montmitonnet, P.; Bodadilla, C.; Foissey, S. The optimal die semi-angle concept in wire drawing, examined using automatic optimization techniques. *Int. J. Mater. Form.* **2013**, *6*, 377–389. [CrossRef]
4. Kabayama, L.K.; Taguchi, S.P.; Martinez, G.A.S. The influence of die geometry on stress distribution by experimental and FEM simulation on electrolytic copper wire drawing. *Mater. Res.* **2009**, *12*, 281–285. [CrossRef]
5. Asakawa, M. "MORDICA LECTURE"—Part 1: Trends in drawing technology for bars and wires. *Wire J. Int.* **2014**, *8*, 60–66.
6. Suliga, M. The Analysis of the High Speed Wire Drawing Process of High Carbon Steel Wires Under Hydrodynamic Lubrication Conditions. *Arch. Metall. Mater.* **2015**, *1*, 403–408. [CrossRef]

7. Golis, B.; Knap, F.; Pilarczyk, J. *Selected Issues from the Theory and Practice of Drawing*; Częstochowa University of Technology scripts, part 6; Wydawnictwo Politechniki Częstochowskiej: Częstochowa, Poland, 1997.
8. Vega, G. Optimization of Shaping by Wire Drawing: Experimental Approach, Modeling and Numerical Simulation. Ph.D. Thesis, Mécanique: Lille University, Lille, France, 7 December 2009.
9. Kania, H.; Liberski, P. Synergistic Influence of the Addition of Al, Ni and Pb to a Zinc Bath upon Growth Kinetics and Structure of Coatings. *Solid State Phenom.* **2014**, *212*, 115–120. [CrossRef]
10. Tatarek, A.; Saternus, M. Research of diffusive dissolution phenomena of reactive steels in zinc bath with bismuth addition. *Corros. Prot.* **2018**, *61*, 186–190.
11. Kania, H.; Skupińska, A. Structures of coatings obtained in a $ZnAl_{23}Mg_3Si_{0.4}$ bath by the batch hot dip method. *Kovove Mater.* **2017**, *55*, 105–111. [CrossRef]
12. Kania, H. Kinetics of Growth and Structure of Coatings Obtained on Sandelin Steels in the High-Temperature Galvanizing. *Solid State Phenom.* **2014**, *212*, 127–132. [CrossRef]
13. Suliga, M. Analysis of the heating of steel wires during high speed multipass drawing process. *Arch. Metall. Mater.* **2014**, *59*, 1475–1480. [CrossRef]
14. Golis, B. *Methods of Assessing Selected Properties of Galvanized and Non-Galvanized Linen Wires*; no. 2; Center for Research and Development of Metal Industry: Kraków, Poland, 1984; Metal Products; pp. 1–30.
15. Golis, B.; Pilarczyk, J.W.; Włudzik, R. *Selected Issues Concerning Wire Processing*; The Wire Association International, Inc. Poland Chapter: Częstochowa, Poland, 2017.
16. Knap, F.; Kruzel, R.; Cieślak, Ł. *Drawing of Wires, Rods and Pipes*; Series Metallurgy No. 36; Częstochowa University of Technology: Częstochowa, Poland, 2004.

MDPI
St. Alban-Anlage 66
4052 Basel
Switzerland
www.mdpi.com

Materials Editorial Office
E-mail: materials@mdpi.com
www.mdpi.com/journal/materials

www.ingramcontent.com/pod-product-compliance
Lightning Source LLC
LaVergne TN
LVHW070632100526
838202LV00012B/785